DUXBURY

# Doing Data Analysis

# with Minitab 14

D0074159

## Robert H. Carver
*Stonehill College*

THOMSON

™

BROOKS/COLE

Australia • Canada • Mexico • Singapore • Spain
United Kingdom • United States

**THOMSON**

**BROOKS/COLE**

Executive Editor: Curt Hinrichs
Assistant Editor: Ann Day
Editorial Assistant: Katherine Brayton
Technology Project Manager: Burke Taft
Marketing Manager: Joseph Rogove
Marketing Assistant: Jessica Perry
Advertising Project Manager: Nathaniel Michelson
Project Manager, Editorial Production: Belinda Krohmer

Print/Media Buyer: Jessica Reed
Permissions Editor: Kiely Sexton
Cover Designer: Denise Davidson
Cover Image: Digital Vision
Cover Printer: Webcom Limited
Printer: Webcom Limited

Printed in Canada
1  2  3  4  5  6  7  07  06  05  04  03

For more information about our products, contact us at:
Thomson Learning Academic Resource Center
1-800-423-0563
For permission to use material from this text, contact us by:
Phone: 1-800-730-2214
Fax: 1-800-730-2215
Web: http://www.thomsonrights.com

Library of Congress Control Number:
2003108609

ISBN 0-534-42084-2

Asia
Thomson Learning
5 Shenton Way #01-01
UIC Building
Singapore 068808

Australia/New Zealand
Thomson Learning
102 Dodds Street
Southbank, Victoria 3006
Australia

Canada
Nelson
1120 Birchmount Road
Toronto, Ontario M1K 5G4
Canada

Europe/Middle East/Africa
Thomson Learning
High Holborn House
50/51 Bedford Row
London WC1R 4LR
United Kingdom

Latin America
Thomson Learning
Seneca, 53
Colonia Polanco
11560 Mexico D.F.
Mexico

Spain/Portugal
Paraninfo
Calle/Magallanes, 25
28015 Madrid, Spain

*For Barry and Cheryl*

# Contents

# Preface

## *Goal of this Book*

This book is a companion for the newcomer to statistical analysis and statistical reasoning. In time, you may come to rely on it as a traveler might rely on a guidebook. Through a series of structured hands-on sessions, this book provides a broad-based introduction to a rich and rewarding field of study.

Unlike a guided tour in which the traveler is a passive spectator, this book leads you on an interactive experience. Our primary vehicle is Minitab 14, and our itinerary is defined by those topics typically encountered in an introductory statistics course. The philosophy of the book is simple: one learns by doing. The sessions in this book present problems, supply real data that relate to those problems, and engage you in doing data analysis to solve those problems.

## *The Changing Environment of Statistics Education*

In the past decade or so, educators have come to reconsider the best approach to teaching and learning in applied statistics courses. With the widespread availability of personal computers, advances in statistical software, and the application of quantitative methods in many professions, courses now emphasize statistical reasoning more than computational skill development. Questions of how have given way to more challenging questions of why, when, and what?

Simultaneously, undergraduates are increasingly comfortable with software, expecting to use computers in their work. Colleges are seeking ways to integrate information technology efficiently into coursework. The introductory statistics course is an ideal place to augment or even replace traditional outside-of-class assignments with structured computer exercises.

Though this book is intended as a supplement an introductory undergraduate statistics course, its predecessor (*Doing Data Analysis with Minitab 12*) has been used in graduate courses and corporate training programs. In all of these settings, students can work independently, learning the software skills outside of class, while coming to understand the underlying statistical concepts and techniques. Instructors can teach statistics and statistical reasoning, rather than algebra or software.

## *The Approach of this Book*

The book reflects the changes described above in several ways. First, and most obviously, it provides instruction in the use of a powerful software package to relieve students of computational drudgery. Second, each exercise is designed to address a statistical issue or need, rather than to feature a particular command or menu in the software. Third, all but one of the datasets in the book is real, representing a variety of disciplines. Fourth, the exercises follow a traditional sequence, making the book compatible with many texts. Finally, as each exercise leads the student through the techniques, it also includes thought-provoking questions and challenges, engaging the student in the processes of statistical reasoning. In designing the sessions, I kept four ideas in mind:

- *Statistical reasoning, not computation, is the goal of the course.* This book asks students questions throughout, balancing software instruction with reflection on the meaning of results.

- *Students arrive in the course ready to learn statistical reasoning.* They need not slog all the way through descriptive techniques before encountering the concept of inference. The exercises invite students to think about inferences from the start, and the questions grow in sophistication as students master new material.

- *Exploration of real data is preferable to artificial datasets.* With the exception of the famous Anscombe regression dataset, all of the datasets are real. Some are very old and some are quite current, and they cover a wide range of substantive areas.

- *Statistical topics, rather than software features, should drive the design of each lab exercise.* Each session features several

Minitab functions selected for their relevance to the statistical concept under consideration.

This book provides a rigorous but limited introduction to the software. Minitab is rich in features and options; this book makes no attempt to "cover" the entire package. Instead, the level of coverage is commensurate with an introductory course. There may be many ways to perform a given task in Minitab; generally, I show one way. This book provides a "foot in the door." Interested students and other users can explore the software possibilities via the extensive Help system or other standard Minitab documentation.

## New in this Edition

Readers familiar with the earlier edition of this book may notice several important changes. Generally, these fall into four categories, as follows:

- *Expanded coverage.* There are two entirely new sessions, covering Two-Way Analysis of Variance (Session 15) and Design of Experiments (Session 22). Coverage of other topics, including conditional probability and normal probability plots has also been augmented in this edition.
- *New Minitab features.* Release 14 of Minitab introduces new graphing interfaces and styles that genuinely facilitate data visualization for the student. Most sessions take advantage of these new features.
- *Improved pedagogy.* Experience with the first edition has suggested some changes in presentation, mostly in the name of clarity. The most noteworthy of these changes is the reorganization of the Sessions introducing one-sample inference. In this edition, Session 10 deals with both confidence intervals and hypothesis tests for a population mean, and Session 11 provides similar coverage for a population proportion.
- *Datasets.* This edition features 67 datasets, of which 46 are new. All of the datasets are compact enough to be fully compatible with both the Student and Professional releases of Minitab 14.

## The Datasets

Each of the datasets provided with this book contains real data, much of it downloaded from public sites on the World Wide Web.

Appendix A describes each file, and each worksheet file contains documentation of sources and detailed variable definitions.

The data files were chosen to represent a variety of interests and fields, and to illustrate specific statistical concepts or techniques. No doubt, each instructor will have some favorite datasets that can be used with these exercises. Most textbooks provide datasets as well. For some tips on converting other datasets for use with Minitab, see Appendix B.

## *Acknowledgments*

Colleagues at Stonehill College and Babson College, as well as participants in the New England Isolated Statisticians group, have shaped my thinking about better ways to teach statistics, have supplied or suggested datasets and examples, and have commented on drafts of these sessions. In particular, I thank Craig Binney, Ralph Bravaco, George Cobb, Roger Denome, Karen Hunt, John McKenzie, Annie Puciloski, George Recck, Norton Starr, and Joan Weinstein. In a class by herself, thanks to Jane Gradwohl Nash for her collegiality and her deep insights and guidance in reshaping several sessions.

Many students through the years have lent a hand to the sessions and datasets in this book and its predecessor. Most recently, Lauren Burokas, Brian Donahue, Kristin Easttey, Kendra Lonquist, Jennifer Lovett, J.P. Mancuso, Tessa Packard, Nick Pontacoloni, and Ryan Spence have all improved the work in a variety of ways. They follow in the footsteps of Jamie Annino, Jason Boyd, Debra Elliott, Nicole Lecuyer, and Bevin Ronayne.

Through the Stonehill Undergraduate Research Experience (SURE) program, I have received considerable support through the years. In particular, Jennifer Karp's service as a SURE scholar ensured the accuracy of the instructors' solutions and other ancillary materials. SURE is but aspect of the institutional support provided by Stonehill College. Bonnie Troupe and Kathy Conroy of the College's Office of Academic Development have consistently delivered competent and congenial aid. Within the Department of Business Administration, Julie Pick has contributed thoughtful, patient, and serene assistance with a variety of document management tasks. Stonehill's Academic Administration, in the persons of Katie Conboy and Karen Talentino, have unwaveringly supplied encouragement and time to bring this project to completion.

At Minitab, Christine Sarris, Gwen Stimely, and Kelli Wilson have answered each of my many questions, fielded numerous requests, and offered timely assistance throughout. I am most appreciative of their attention and good graces.

Once again, I am deeply grateful to everyone at Duxbury Press for their work on this and earlier projects. Curt Hinrichs, Ann Day, and Joe Rogove have seen this manuscript through its iterations, and have been responsive and extremely helpful. Professor Bruce Trumbo, California State University, scrutinized each page making numerous improvements throughout, as well as serving as a reviewer for the first edition. Thanks also to the following reviewers whose constructive suggestions have improved the quality of the first edition: Gregory Davis, University of Wisconsin, Green Bay; Robert Fountain, Portland State University; Jeffrey Jarrett, University of Rhode Island; Dennis Kimzey, Rogue Community College; and Roxy Peck, California Polytechnic State University. The remaining flaws are all mine.

Finally, I thank my family: Donna, Sam and Ben. All too often, I have had to borrow time and attention from them to bring this project to a timely conclusion.

## To the Student

This main goal of this book is to help you understand the concepts and techniques of statistical analysis. It can supplement but not replace your principal textbook or your classroom time. To get the maximum benefit from the book, you should take your time and work carefully. Read through a session before you sit down at the computer. Each session should require no more than about 30 minutes of computer time; there's little need to rush through them.

You'll often see numbered questions interspersed through the computer instructions. These are intended to shift your focus from 'getting answers' to thinking about what the answers mean, whether they make sense, whether they surprise or puzzle you, or how they relate to what you have been doing in class. Attend to these questions, even when you aren't sure of their purpose.

As noted earlier, Minitab is a large and powerful software package, with many capabilities. Many of the features of the program are beyond the scope of an introductory course, and do not figure in these sessions. However, if you are curious or adventurous, you should explore the menus and Help system. You may find a quicker, more intuitive, or more interesting way to approach a problem.

Each session ends with a section called *"Moving On...."* You should also respond to the numbered questions in that section, as assigned by your instructor.

# About the Author

Robert H. Carver is Chair and Professor of Business Administration at Stonehill College in Easton, Massachusetts where his teaching has been recognized with the College's annual Excellence in Teaching award. In addition to Business Statistics, he teaches courses in information systems, business and society, and the Department's Senior honors research seminar. Professor Carver serves on the Board of the Journal of Statistics Education.

He is the author of *Doing Data Analysis with Minitab 12* and (with Jane Gradwohl Nash) *Doing Data Analysis with SPSS 10.0* (Duxbury Press). His work has appeared in *Publius, The Journal of Statistics Education, PS: Political Science & Politics, Public Administration Review, Public Productivity Review,* and *The Journal of Consumer Marketing.* He holds an A.B. from Amherst College and a Ph.D. in Public Policy from the University of Michigan.

# A First Look at Minitab

## *Objectives*

In this session, you will learn to:

- Launch and exit Minitab
- Enter quantitative and qualitative data in a worksheet
- Create and print a graph
- Get Help
- Save your work to a disk

## *Launching Minitab*

Click and hold the left mouse button on the 🏁 **Start** button at the lower left of your screen, and drag the cursor to select Programs. Locate and choose **MINITAB 14**. Move the cursor to the right, select **MINITAB 14**, and click the left mouse button to launch the program.

Because Minitab is a large program, you may have to wait a few moments before the program is ready for use. On the next page is an image of the screen you will see when Minitab is ready. There is a menu bar across the top of the screen and two open windows. The upper window is the *Session Window*, which will contain the results of all commands you issue to Minitab. The lower window is the *Data Window*, which is used to display the data that you will analyze using the program. Later, you will see that it is possible to have several open data windows, showing multiple worksheets. For now, we'll work with just one. Each window has a unique purpose, and each can be saved separately to disk. It's important at the outset to have a sense of what each window is about. We'll get to the details about each type of window later.

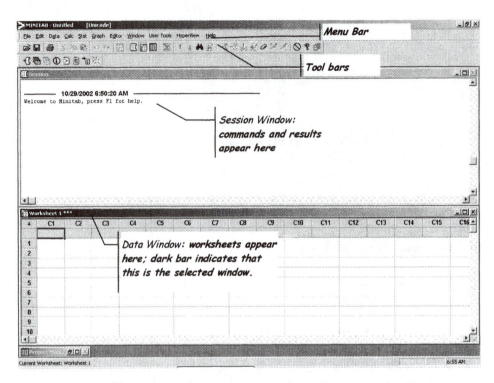

At any point in your session, only one of the Windows is *selected*, meaning that mouse actions and keystrokes will affect that window alone. To select a window, place the cursor anywhere in the window, and click the left button once. When you start Minitab, the Data Window is initially selected.

Since Minitab operates upon data, we usually start by placing data into a worksheet, either from the keyboard or from a stored disk file. The Data Window looks much like a spreadsheet. Cells may contain numbers or text, but unlike a spreadsheet, they never contain formulas. Columns are numbered C1, C2, and so on. Except for the first row, which is reserved for variables names, rows are numbered consecutively. Each variable in your dataset will occupy one column of the Minitab worksheet, and each row represents one observation. For example, if you have a sample of fifty observations on two variables, your worksheet would contain two columns and fifty rows.

There is another window that is minimized when Minitab starts, so you do not see it: the *Project Manager* Window. As you progress through a single session with Minitab, you may create several graphs or generate several analyses. The collection of all of worksheets, graphs and session output are known as a *project*. The Project Manager allows you to navigate through the various windows and files in your Minitab project. It also preserves a complete record of your work in a given session. We'll work with the Project Manager in future sessions.

You can also preserve a record of all commands right within the Session Window itself. By default, Minitab only shows the results of computations or commands in the Session Window. As you are learning Minitab, it is a good idea to have a record of your keystrokes and commands during each session. To change the default setting, follow these instructions[1] :

⌐ Click anywhere in the Session Window to make it the active (or selected) window.

⌐ On the **Editor** menu, select **Enable Commands** as shown here, and click.

You may notice that the notation MTB> appears in the Session Window. This indicates that you have successfully enabled the Command Language feature. From now on, any menu commands you select will be recorded in the Session Window. Moreover, you may now enter some commands directly, without using menus.

As a first command, do this:

⌐ In the session Window, type **note {your name}'s First Session**, substituting your own name where indicated. The word "note" is actually a Minitab command that allows a user to add comments to the session output.

---

[1] This is the only time you'll need to follow these instructions.

As you use Minitab, other windows may open as you create graphs or issue various commands. Any window can be enlarged, shrunk, or closed altogether by clicking on the small icon boxes in the upper right corner of the window, or by clicking and dragging the lower right corner of the window.

The menu bar across the top of the screen identifies categories of Minitab's features. There are two ways to issue commands in Minitab: choosing commands from the menu or icon bars, or typing them directly into the Session Window using Session commands.[2] This book almost always refers you to the menus. As you make menu choices, Minitab actually "writes" a command into the Session Window and into a "History" folder, so even with menus you are actually using Session Commands. You can do no harm by clicking on a menu and reading the choices available, and you should expect to spend some time exploring your choices in this way. As you develop your skills and become more familiar with the software, you may want to use icons or Session Commands.

## *Entering Data into a Worksheet*

For most of the sessions in this book, you will start by accessing data already stored on a disk. For small datasets, though, it is easy to type in the data. For this session, you will transfer the data displayed below into the empty worksheet in the Data Window. This will give you a chance to see how the data move from the real world into a worksheet.

In this first example, we will create a worksheet, and then use Minitab to construct two graphs from the data. We will save our entered data into a Minitab *Worksheet file*. This is typical of the tasks you will perform throughout the book.

This small set of data comes from a classroom experiment.[3]  The class was divided into teams, and each team received specifications for the construction of a paper helicopter. A paper helicopter is a narrow strip of paper that has been cut and folded in the shape of a helicopter propeller. When released from a height, it spins gracefully to the ground. The goal of the experiment is to identify some variations in helicopter

---

[2] It visibly writes the command in the Session Window only if you have enabled the command language, as described earlier.

[3] This experiment was inspired by G.E.P. Box and P.Y.T. Liu (1999) "Statistics as a catalyst to learning by scientific method part I—an example." *Journal of Quality Technology* 31(1), pp. 1–15. Original dimensions used in this experiment were found at www.family.com under "paper helicopters."

design that will tend to increase the flight duration of the helicopters. The basic design is illustrated below.

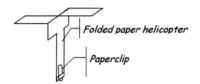

Folded paper helicopter

Paperclip

Each team was instructed to build and test-fly a slightly different helicopter design, timing the duration of eight repeated flights. In this first exercise, we'll study the flight times for two of the designs.

The difference between the two helicopter designs was that one included a paper clip and the other did not, having only a folded lower body. The durations of the eight trials were as follows:

| | Flight Trial (time in seconds) | | | | | | | |
|---|---|---|---|---|---|---|---|---|
| Design | 1 | 2 | 3 | 4 | 5 | 6 | 7 | 8 |
| Clip | 3.3 | 3.4 | 3.5 | 3.6 | 3.4 | 3.6 | 3.5 | 3.3 |
| Fold | 3.6 | 4.4 | 4.9 | 4.5 | 4.5 | 4.5 | 4.1 | 4.6 |

First we will enter these recorded times into a Minitab worksheet. There are several ways we might organize the data in the worksheet; see Appendix C for a full discussion of these options. We will almost always place the data for each variable in a different column. In this example, we have two variables: the design of the helicopter, and the length of its flight. In all, the students "observed" each of the variables repeatedly, for a total of 16 observations.

The first row in a worksheet, above the numbered row 1, is reserved for variable names. Although Minitab does not require us to use variable names, we will do so for our convenience. In the absence of a name, each variable is simply identified by the column number: C1, C2, etc.

Move the cursor into the Data Window, and position it in the cell below the heading **C1**. Click the left mouse button once, and type **Design**; press **Enter**. The cursor will move down into Row 1. Type **Clip**, and press **Enter** (see illustration next page). This represents the fact that the first observed flight was used a helicopter with a paper clip. Now return to that first cell and position the cursor over the small black square in the lower right corner of the cell, until the cursor symbol changes to a black cross. Click and drag downward until you highlight the cells in rows 1 through 8, then release the mouse button. This copies the word "Clip" into each of the highlighted cells.

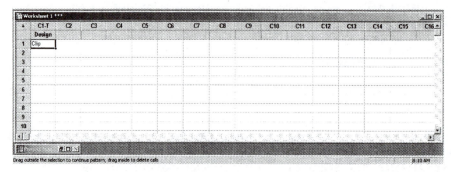

Note that the **C1** at the top of the column has changed to **C1-T**, indicating that this column refer to a text (i.e., qualitative) variable.

🖰  Continue by typing **Fold** into the 9th row cell; use the same technique as above to copy the word Fold into rows 10 through 16.

🖰  Title the second variable **Duration**, and enter the data from the previous page into Rows 1 through 16 within the second column, and then maximize your Data Window. It should look like this:

| | C1-T | C2 | C3 | C4 | C5 | C6 | C7 | C8 | C9 | C10 | C11 | C12 | C13 | C14 | C15 | C16 |
|---|---|---|---|---|---|---|---|---|---|---|---|---|---|---|---|---|
| | Design | Duration | | | | | | | | | | | | | | |
| 1 | Clip | 3.3 | | | | | | | | | | | | | | |
| 2 | Clip | 3.4 | | | | | | | | | | | | | | |
| 3 | Clip | 3.5 | | | | | | | | | | | | | | |
| 4 | Clip | 3.6 | | | | | | | | | | | | | | |
| 5 | Clip | 3.4 | | | | | | | | | | | | | | |
| 6 | Clip | 3.6 | | | | | | | | | | | | | | |
| 7 | Clip | 3.5 | | | | | | | | | | | | | | |
| 8 | Clip | 3.3 | | | | | | | | | | | | | | |
| 9 | Fold | 3.6 | | | | | | | | | | | | | | |
| 10 | Fold | 4.4 | | | | | | | | | | | | | | |
| 11 | Fold | 4.9 | | | | | | | | | | | | | | |
| 12 | Fold | 4.5 | | | | | | | | | | | | | | |
| 13 | Fold | 4.5 | | | | | | | | | | | | | | |
| 14 | Fold | 4.5 | | | | | | | | | | | | | | |
| 15 | Fold | 4.1 | | | | | | | | | | | | | | |
| 16 | Fold | 4.6 | | | | | | | | | | | | | | |
| 17 | | | | | | | | | | | | | | | | |
| 18 | | | | | | | | | | | | | | | | |
| 19 | | | | | | | | | | | | | | | | |

## Saving a Worksheet

It is wise to save all of your work in a disk file. Minitab distinguishes among several objects that one might want to save: a Session Window, a worksheet, a graph, or a collection of related items, which Minitab refers to as a *Project*. For further discussion and description of these options, consult Appendix B. At this point, we've created a worksheet and ought to save it on a diskette. Let's call the worksheet Copter.

> 💻 Check with your instructor to see if you can save the worksheet on a hard drive or network drive in your system.

🖱 On the **File** menu, choose **Save Current Worksheet As...** In the **Save in** box, select **3 ½ Floppy (A:)**. Then, next to File Name, type **Copter**.

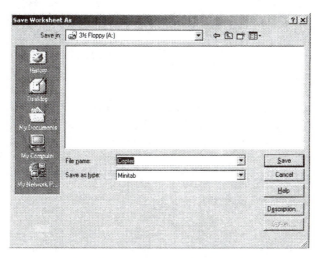

This *just* saves the data in the worksheet. Your Session and other windows have *not* been saved. That is a skill to be covered soon.

## *Exploring the Data*

It is always good practice to graph a set of data, to look for obvious errors in data entry, to notice striking patterns, and to familiarize oneself with the dataset before further investigations. One with numeric data like flight times, one simple exploratory tool is the *dotplot*. We'll study dotplots more fully in Session 2, but let's create a simple one now.

🖱 Click on **Graph** in the menu bar, and choose **Dotplot**. This initiates a short sequence of dialog boxes.

🖱 As shown below left, we first indicate that we are creating a simple dotplot for one variable. This opens the dialog shown to the right.

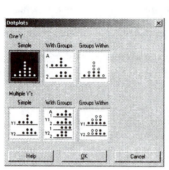

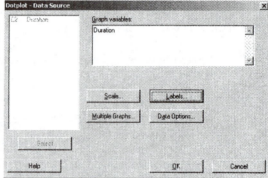

🖱 Move the cursor into the box in the upper left of the dialog, and highlight **C2 Duration**. Then click on the **Select** button, or double-click over **C2 Duration**. You'll see **Duration** appear in the box marked **Variables**. Note that C1 is missing from the list; a dotplot is for numerical data, and C1 contains text data.

🖱 Move the cursor to the button marked **Labels...** and click, opening the dialog box shown below. Type **Flight Durations** and your name as shown. You should always add a descriptive title to graphs you create, to credit yourself for your work, and to cite sources of data when appropriate.

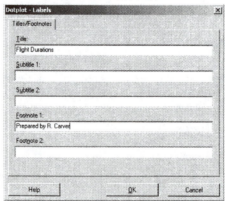

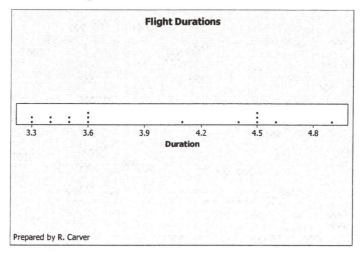 Now click **OK**.

You will see a new window appear, containing a dotplot (see below). This graph displays the sixteen data values that you entered into the worksheet. The horizontal axis represents a single flight. If more than one dot appears above the horizontal axis, that just means that more than one flight lasted that long. For example, two flights lasted 3.4 seconds, and three flights lasted 3.6 seconds.

**Flight Durations**

Prepared by R. Carver

1. *What strikes you as noteworthy about this graph?*

2. *What does it tell you about how the paperclip affects flight durations?*

In fact this graph tells us nothing relevant to the second question because the graph does not represent the design variable in our worksheet. To answer the second question, we must somehow create a plot that compares the flight times for the two designs. We'll do so by making two dotplots that share the same horizontal axis.

3. *Before doing so, take a moment to think about what you expect this graph to look like, having seen the other one. How do you expect the two plots to compare?*

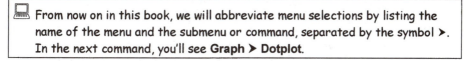 From now on in this book, we will abbreviate menu selections by listing the name of the menu and the submenu or command, separated by the symbol ➤. In the next command, you'll see **Graph ➤ Dotplot**.

🖰 **Graph ➤ Dotplot...** In the first dialog, choose With Groups, as shown.

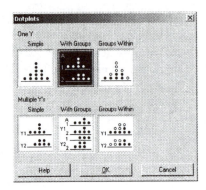

🖰 Notice that the next dialog box is a bit different from before. Complete the dialog as shown below, and use the Labels dialog to add a title and footnote again.

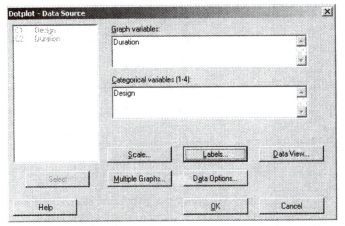

4. *Look at your new graph. How does it compare to the first one?*

5. *Which helicopter design seems to lead to longer flights? Explain your thinking.*

## Saving a Project

At this point, we have two open graphs, a Session Window with some commands, and a Data Window with a worksheet. We could save each window in a separate disk file before ending this session. Alternatively, we can save them all as a Minitab *Project File*. As noted earlier, a project is a related collection of worksheets and other windows. To save a project, do the following:

🖰 **File ➤ Save Project As...**In this dialog, assign a name to the file (**Session 1**). This new file will save all of the windows you see, as well as those that you don't see.

## *Getting Help*

You may have noticed the **Help** button in a dialog box. Minitab features an extensive on-line Help system. If you aren't sure what a term in the dialog means, or how to interpret the results of a command, click on **Help**. You can also search for help on a variety of topics via the **Help** menu at the top of your screen. As you work your way through the sessions in this book, this feature may often be valuable. Spend some time experimenting with it before you genuinely need it.

## *Printing in Minitab*

Now that you have created some graphs, let's print one of them.

🖰 To print a graph, first select the graph of interest by clicking somewhere on the graph window.

---
💻 If you are using a campus public computer, check with your instructor about special considerations in selecting a printer or issuing a print command.
---

🖰 **File ➤ Print Graph...**This command will print the selected window. Click **OK**.

In general, to print all or part of a window, you must make the window active by clicking on it, and then find the appropriate print command in the **File** menu.

## *Quitting Minitab*

When you have completed your work, it is important to exit the program properly. Virtually all Windows programs follow the same method of quitting.

🖰 **File ➤ Exit** You will see a message asking if you wish to save changes to this project. Since we saved everything earlier, click **No**.

That's all there is to it. Later sessions will explain menus and commands in greater detail. This session is intended as a first look; you will return to these commands and others at a later time.

## *Further Reading*

Box, G.E.P. and Liu, P.Y.T. (1999). "Statistics as a catalyst to learning by scientific method part I—an example." *Journal of Quality Technology* 31(1), pp. 1–15.

# Tables and Graphs for One Variable

## *Objectives*

In this session, you will learn to:
- Retrieve data stored in a Minitab worksheet
- Inspect the contents of a worksheet
- Understand the difference between quantitative and qualitative data
- Explore quantitative data with a Dotplot
- Explore quantitative data with a Stem-and-Leaf display
- Create and customize a histogram
- Create a frequency distribution
- Print output from the Session Window
- Create a bar chart for qualitative data
- Understand the difference between cross-sectional and time-series data
- Plot data collected over time

## *Opening a Worksheet*

In the previous session, you created a Minitab worksheet by entering data into the Data Window. In this lab, you'll use several worksheets that are available on your diskette. This session begins with some detailed data about highway travel and safety in the United States.

> ⌨ NOTE: In this chapter, we'll assume that the *Doing Data Analysis* files are on your C: drive. The location of Minitab files might be different on your computer system. If you have a problem, check with your instructor.

🖱 Choose **File ➤ Open Worksheet....** A dialog box like the one shown on the next page will open. Select the drive and directory containing the files that accompany this book, and you will see a list of available worksheet files. Scroll to the right and select the one named **StateTrans**.

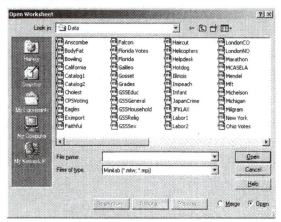

Click **Open**, and you may see the message shown below, alerting you that your current project will be augmented with this file. If you do, click **OK**, and your Data Window will show the data from the worksheet file. Using the *scroll bars* at the bottom and right side of the screen, move around the worksheet, just to look at the data. Appendix A of this text provides the sources of the data found in these worksheets. The next section explains how you can obtain more information about these data files during your Minitab session.

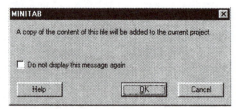

## *The Project Manager Window*

After you click OK, the worksheet will appear in the data window on your screen, and should look like the figure on the next page. This worksheet contains fourteen different variables, with observations for each of the 50 states plus the District of Columbia during 1998.

⁀ᗺ Use the scroll bars to inspect the different variables and data values.

Some of the variable names obviously indicate the meaning of the variable, but some are obscure. To find out a little more about what is in this worksheet, we want to consult the information in the worksheet description and the variable descriptions contained in the *Info* window.

⁀ᗺ **Window ➤ Project Manager** This command opens the Project Manager window, which allows you to keep track of the various worksheets, graphs, and other files that you create during a session.

In the left pane of the window, you'll find a tree structure showing all of the folders and documents open in this project session. We have opened StateTrans as our first worksheet.

⁀ᗺ Click on the folder icon labeled **StateTrans.MTW**. In the right pane, you will see a general description of the worksheet.

⁀ᗺ Now click on the subfolder marked **Columns**. You will find an extensive description of the contents of the worksheet.

Specifically, you will see a list of all variable names, the columns they occupy, the number of available observations, the type of data (*T* for text or qualitative; *N* for numeric; *D* for dates), and a description of the meaning of each variable. In our first analysis, we'll focus on the variable in **C7 AVMTPC**, which represents the average number of miles driven *per capita* in 1998.

## *A Dotplot*

A Dotplot is a good tool for a first look at the shape and spread of quantitative data.

🖱 Choose **Graph ➤ Dotplot…**. As in the prior session, we'll choose a simple dotplot. In the main dialog box, move your cursor to the list of variables on the left of the box, and click on **AVMTPC**. Then click on **Select**. You will see the variable name **AVMTPC** appear in the **Variables** box.

🖱 Click on **Labels** and add a descriptive title, crediting yourself as the creator of the graph, and click **OK**.

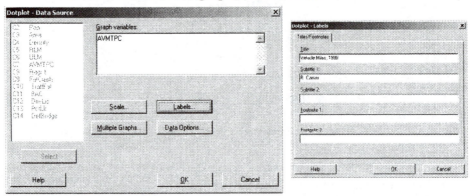

After clicking **OK**, another window (shown facing) will appear. In this plot, the horizontal axis represents a number of per capita vehicle miles, and the vertical axis represents the number of states reporting that many miles. Each dot in the graph represents one state.

Notice that the values vary considerably, but that many states report annual vehicle travel in the range of about 9000 to 12000 miles.

1. *How would you describe the shape of this distribution?*

2. *Also notice the one dot at the right end. What does it mean?*

3. *What state might it represent?  See if you can locate that one outlier in the Data Window.*[1]

---

[1] Suggestion: On Minitab's Help menu, search for Help on "Brush" to learn how to point to an outlier and locate its value in the data window.

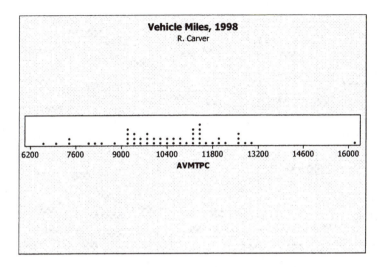

## *Exploring the Data with Stem-and-Leaf*

Another simple tool for exploring data is a Stem-and-Leaf plot. The Stem-and-Leaf display is a tool for developing meaningful frequency distributions, and provides a simple visual display of the data.

🖰 **Graph ➤ Stem-and-Leaf...**[2] Select **AVMTPC**, de-select **Trim outliers**, then click **OK**; the Stem-and-Leaf diagram will be in the Session Window. You may want to *maximize* the Session Window for a better look.

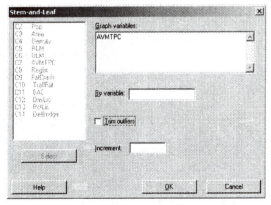

---

[2] This command is also available under **Stat ➤ EDA ➤ Stem-and-Leaf**.

In this output, we see first that there are 51 observations, and that each stem entry represents 100 miles. In the display, there are three columns of information, representing *cumulative frequency, stems,* and *leaves.* Let's consider the meaning of each column.

The first column shows cumulative frequency up to the median, and then shows declining cumulative frequency. In this example, the median—denoted by parentheses around the value—occurs in the fifth class of the data. Below that, cumulative frequencies "count down."

Recall from the dotplot that these data values range from about 6,600 through 16,200. In a stem and leaf plot, we think of a data value as being composed of two parts. A *stem* is the first one or two significant digits, and a *leaf* is one digit beyond the stem. Since these data values are all in thousands, the stems are the values to the left of the comma.

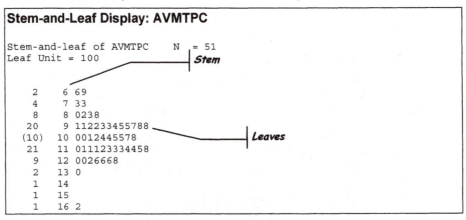

**Stem-and-Leaf Display: AVMTPC**

```
Stem-and-leaf of AVMTPC     N = 51        Stem
Leaf Unit = 100

     2     6  69
     4     7  33
     8     8  0238
    20     9  112233455788
   (10)   10  0012445578                  Leaves
    21    11  011123334458
     9    12  0026668
     2    13  0
     1    14
     1    15
     1    16  2
```

This first row in the diagram shows that drivers in 2 states drive, on average, fewer than 7000 miles annually; in one, they drive approximately 6600 miles and in the other, they drive approximately 6900 miles. From the second row of the display, we find that two more states report per capita mileage in the 7000–7999 mile range; as it happens both have mileage figures of approximately 7300 miles. In all, four states report annual per capita mileage under 8000 miles. In the next row, one state reported about 8000 miles, one had about 8200, another 8300, and one more had about 8800.

**4.    Look at the row with a stem value of 13. Explain the meaning of the digits in that row.**

Skip down to the stem value of 14; there are no leaves, meaning that there were no states in the 14,000 – 14,999 mile range.

## Creating a Histogram

A *histogram* is a visual display of a frequency distribution for quantitative data. The horizontal axis represents the possible values of the selected variable, and the vertical axis represents frequency.

**Graph ➤ Histogram...** Choose a simple histogram, and select the variable **AVMTPC**. Click on **Labels** and type a title for this graph (e.g., "1998 Annual Vehicle Miles per Capita") and add your name as a subtitle or footnote. Click **OK** and then click **OK** again, and a new window will open, containing the graph.

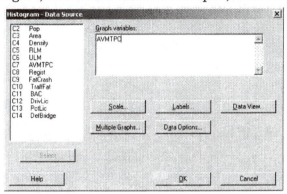

5.    *Compare this histogram to the dotplot and stem-and-leaf displays. What differences, if any, do you see?*

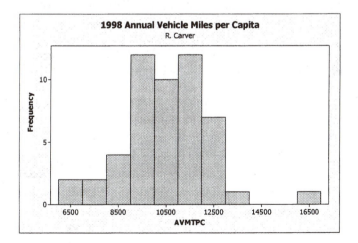

In this histogram, Minitab determined the number of bars, which affects the apparent shape of the distribution. We can control the number of bars as follows:

🖱 In the graph window, move your cursor over one of the numbers along the horizontal axis, and double click. This brings up a new dialog box; select the tab marked **Binning**.

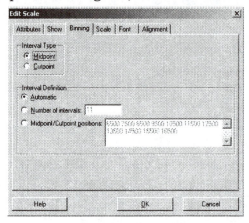

🖱 Change the **Interval Type** from **Midpoint** to **Cutpoint**. This resets the relationship of the bars to the horizontal axis, so that the endpoints of class intervals are labeled.

🖱 Click the button next to **Number of Intervals** and type **8** in the box. Click **OK** to update the histogram.

6. *How does this compare to your first one?*

7. *Which graph better summarizes the dataset? How so?*

Sometimes we can get a clearer sense of the general shape of a dataset by altering the number of intervals. Experiment with several different values, noting the impact of the changes. Here we have a distribution with one central peak at approximately 10,000 miles, and the distribution is generally *symmetrical* on either side of 10,000 miles.

8. *What does the peak at 10,000 miles tell you?*

9. *What is the relevance of the symmetric pattern in this data?*

Now, let's see a graph of the *cumulative* frequency distribution. Cumulative frequency represents the number of data observations at or below a particular variable value.

🖱 Move your cursor to the vertical axis, and double click on one of the axis values. In the dialog box, select the **Type** tab.

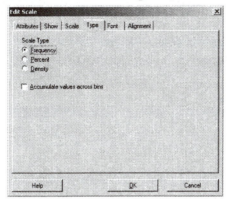

🖱 Check the box marked **Accumulate values across bins**, and click **OK**. This will accumulate the frequencies for each *bin* or group of values along the horizontal axis.

Compare the results of this graph to the prior graph.

**10. *What does this graph show?***

**11. *About how many states report annual per capita mileage of less than 11,250 miles?***

## *Frequency Distributions with Tally*

Let's switch to another worksheet and a set of questions about *qualitative*, or *categorical*, data. One of the major news stories in the summer of 2000 was the serious problem with Firestone tires. Let's have a look at some of the data behind the stories.

🖱 **File ➤ Open Worksheet**. Choose the worksheet called **Tires**.

This file contains a sample of 71 reports of tire failures filed with the U.S. Department of Transportation during the month of July 2000. There are twenty-one variables in the worksheet, representing different characteristics or aspects of each incident of tire failure. The variable **VEH_MFR** occupies C9, and indicates the name of the automobile company that manufactured the vehicle whose tire failed. This is a *categorical*, or *nominal*, variable because it simply names the company.

With a categorical variable, the simplest summary is a list of all of the vehicle manufacturers in the worksheet, along with a tally of their frequency. We can have Minitab generate tally.

🖱 **Stat ▸ Tables ▸ Tally Individual Variables...** In the Tally dialog box, select variable **VEH_MFR** and click **OK**. The tally results appear in the session window.

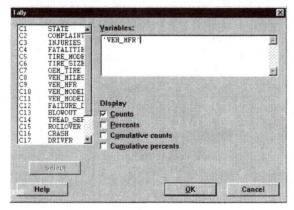

In the session window, you now see this:

### Tally for Discrete Variables: VEH_MFR

| VEH_MFR | Count |
|---|---|
| * | 5 |
| CHEVROLET | 1 |
| FORD | 60 |
| GMC | 1 |
| JEEP | 1 |
| MAZDA | 1 |
| TOYOTA | 2 |
| N= | 71 |

Note: the "VEH_MFR" column shows values of the variable, and the "Count" column shows frequencies.

12. **Which kind of automobile was most often involved in these incidents?**

13. **How does the frequency of that company compare to the other companies?**

## *Printing Session Output*

Sometimes you will want to print all or part of a Session Window. Before printing your session, be sure you have made session output editable (see Session 1), and typed your name into the session. You can print the entire session by clicking anywhere in the Session Window, and then choosing **File ➤ Print Window**. To print *part* of a Session Window, do the following:

➧ In the Session Window, move the "I-beam" cursor to the beginning of the portion you want to print. Click the left mouse button.

➧ Using the scroll bars (if necessary), move the cursor to the end of the portion you want to print. Then press Shift on the keyboard and click the left mouse button. You'll see your selection highlighted.

➧ **File ➤ Print Session Window...** Notice that the Selection button is already marked, meaning that you'll print only the highlighted selection of the Session Window. Click OK.

## *A Bar Chart*

To graph this tally, we should make a bar chart. Like a histogram, a bar chart uses vertical bars to represent frequency. However, unlike a histogram, the values on the horizontal axis represent *categories*, rather than quantitative values. As such, there is no obvious preferred ordering of the bars from left to right.

➧ **Graph ➤ Bar Chart....** Choose a simple bar chart. In the dialog box, simply select the variable **VEH_MFR** once again. As always, you should title and claim authorship for your graph before clicking **OK**.

14. *Compare the chart and the tally. They should contain the same information. Do they?*

15. *What are the relative strengths and weaknesses of each method of displaying this distribution?*

## A Time Series Graph

In the previous examples, each data observation represented many different states or events (accidents) during a single time period. We refer to data gathered in this fashion as *cross-sectional data.* Some datasets are gathered by *repeatedly* observing a single item over multiple regular time periods. We call such datasets *time series,* or *longitudinal* data. When we analyze time series data, we often look for patterns or trends evolving over time.

Though most of the worksheets in this book contain cross-sectional data, several have time series datasets. One of them is called **US**, and the worksheet contains several demographic and economic variables for the United States between the years 1965 and 2000. Open the worksheet now; we'll start by displaying the population in the U.S. during this period.

🖱 **Graph ➤ Time Series Plot...** In the first dialog, choose a simple plot, and then select the variable **Pop**.

As you look at the resulting graph, note that the horizontal axis, labeled **Index**, represents the sequence of observation periods. That is, the population for 1965 is shown at index value 1. The vertical axis represents the population of the U.S. Looking at this graph, it is clear that the population has grown steadily year after year.

From this graph, it is apparent that the story of this variable is one of *growth,* and this is one reason we construct time series plots: Sometimes the variation in a variable follows a predictable pattern. Suppose we had decided to graph this variable using a histogram; let's see what would happen.

🖱 **Graphs ➤ Histogram...** Make a simple histogram of **Pop**.

16. *Explain what the horizontal and vertical axes represent in this graph. Does this graph tell the story of population growth? Explain.*

Some time series variables do not grow or decline steadily, but vary in other ways. Consider the variable called **Unemprt**, which is the annual unemployment rate in the United States.

🖱 Make a time series plot for **Unemprt**.

17. *Describe what you see in this graph; is there a predictable pattern? Explain.*

The graphs presented in this chapter represent a few of the available types. Each is suitable for a different type of data, and for answering different questions. In these illustrations, we have typically used the default settings, but noted that many options are available. In later chapters, we will expand our graphical repertoire, but this chapter introduces several standard approaches to summarizing data visually.

When you look at a graph or table, be sure to orient yourself by determining the meaning of column headings and axis labels. Remember that the purpose of these tools is to summarize and describe. Good graphs tell a story clearly and without clutter.

## Further Reading

Tufte, E. (1983) *Visual Display of Quantitative Information.* Cheshire CT: Graphics Press.

Wainer, H. (1997). *Visual Revelations.* New York: Copernicus.

## Moving On...

Using the skills you have practiced in this session, now answer the following questions. In each case, provide an appropriate graph or table to justify your answer, and explain how you draw your conclusion. You may be able to use several approaches or commands to answer the question; think about which approach seems best to you. The first three questions return to the **Tires** data. Later questions refer to other worksheets. Consult Appendix A and the Minitab Project Manager for detailed descriptions of the worksheets and variables.

### Tires

1. In each of the incidents recorded in the **Tires** worksheet, between one and four tires failed. Create an appropriate graph to display the variable called **NumFAIL**, and comment on what the display shows.

2. The data in columns 17 through 20 indicates which tire(s) failed; which tire position seems to have been the most common location of failures? Which technique did you use to draw your conclusion?

## StateTrans

3. The variable named **BAC** refers to the legal blood alcohol threshold for driving while intoxicated. All states set the threshold at either .08 or .10. About what percentage of states use the .08 standard?

4. The variable called **DefBridge** is the percentage of bridges in the state that were considered deficient and obsolete in 1998. Do all states have similar percentages of bridges needing replacement or repair? What seems to be a typical percentage? How much variation is there across states?

5. The variable called **TraffFat** is the number of motor vehicle accident fatalities during the year 1998. Use a graph to help you determine approximately how many states had more than 1,500 fatalities that year? Explain your choice of graphical tool. (Presumably, this variable is related to population and mileage; in Session 3, we'll see how to take that relationship into account.)

## Triathlon

This file contains the finish times for the Women's Triathlon in the 2000 Olympic games.

6. The variable **Country** is the home country of the racer. Which countries had the most women competing in this event? (Hint: Use a tally or a bar chart)

7. How would you characterize the shape of the histogram of the variable **Minutes**? (Experiment with different numbers of intervals in this graph.)

8. Approximately what percentage of these athletes completed the event in less than 2 hours, 6 minutes (126 minutes)?

## MCASELA

This worksheet contains school-by-school results of the 2000 Massachusetts Comprehensive Assessment System (MCAS) Grade 4 exams for the English Language Arts portion of the exam.

9. Use one of the tools introduced in this session to explore the shape of the variable called **profper**, which is the percent of students scoring at the proficient level. Describe the center, shape, and spread of this distribution.

10. Use one of the tools introduced in this session to explore the shape of the variable called **niper**, which is the percent of students scoring at the "needs improvement" level. Describe the center, shape, and spread of this distribution.

11. Use one of the tools introduced in this session to explore the shape of the variable called **mnscore**, which is the mean (average) score of students at each school. Describe the center, shape, and spread of this distribution.

## Terrorism

This file contains a time series of the number of terrorist incidents recorded by the U.S. Central Intelligence Agency (CIA) between 1971 and 1996.

12. Create a time series plot of the annual number of incidents reported, and describe what this graph shows.

13. Now make a dotplot for the number of incidents, and describe what you see in it.

14. Which graph does a better job of displaying the data, and why?

## GSSRelig

This file contains responses from the 1998 General Social Survey. The data in this worksheet refer to the GSS questions about religious practices and belief.

15. What were the religious affiliations of the respondents in this survey, and which religions were most and least represented among respondents?

16. Create an appropriate visual summary of responses to the question: "Does the Respondent believe in Heaven?" (Hint: check the Info Window for variable descriptions)

17. Create an appropriate visual summary of responses too the question "Does the Respondent believe in Hell." Compare the results to the previous question.

## JapanCrime

This file contains time series data about reported crimes and arrests in Japan between 1984 and 1999.

18. Create a time series plot of the number of reported crimes and describe what you see in the graph.

19. Create a time series plot of the number of arrests and describe what you see in the graph.

20. Create a time series plot of the *ratio* of arrests to reported crimes, and describe what you see in the graph.

21. In comparing arrests from year to year, is it more reasonable to look at the total number or at the ratio of arrests to reported crimes? Explain your reasoning.

## Texas Votes

In the US Presidential election of 2002 George W. Bush, the governor of Texas, was elected President of the United States. This worksheet contains the vote tallies from each county in the state of Texas.

22. Create a visual display showing the number of votes cast for Bush. Comment on the shape, center, and spread of the distribution.

23. Create a visual display showing the number of votes cast for Gore. Comment on the shape, center, and spread of the distribution in comparison to the distribution of votes for Bush.

24. Create a visual display showing the percentage of eligible voter who actually voted in the election. Comment on the shape, center, and spread of the distribution.

## *Notes*

The Moving On... questions in the session require some forethought and may raise questions in your mind. Use these pages to keep track of questions and ideas.

# Tables and Graphs for Two Variables

## *Objectives*

In this session, you will learn to:

- Represent two variables in different ways in a worksheet
- Cross-tabulate two variables
- Create a bar graph comparing two variables
- Create a histogram for two variables
- Create a scatter plot for two quantitative variables
- Create a time-series plot comparing two variables

## *Cross-Tabulating Data*

In the first example, we will consider a case of two qualitative variables. The example uses the General Social Survey data, and looks at the marital status of the people who responded to the survey. Specifically, we'll compare the reported marital status of the women to that of the men. Since marital status and gender are both categorical data, our first approach will be a *joint frequency table* also known as a *cross-tabulation*.

☞ Open the file by selecting **File ➤ Open Worksheet...** and choosing **GSSGeneral**.

☞ **Stat ➤ Tables ➤ Cross Tabulation and Chi –Square...**[1] You will see the dialog box show on the next page. Just select the variables

---

[1] Chi-Square refers to a statistical technique that is covered in Session 13. For now, we are just interested in the cross-tabulation.

MARITAL for rows and SEX for columns. Under Display, check
Counts and click OK. You'll find the cross-tabulation in the
Session Window.

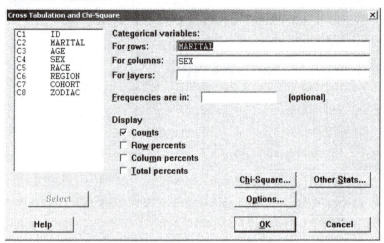

## Tabulated Statistics: MARITAL, SEX

Rows: MARITAL    Columns: SEX

|            | Female | Male | All  |
|------------|--------|------|------|
| Divorced   | 135    | 83   | 218  |
| Married    | 339    | 342  | 681  |
| Never Marr | 169    | 183  | 352  |
| Separated  | 24     | 22   | 46   |
| Widowed    | 124    | 24   | 148  |
| All        | 791    | 654  | 1445 |

### 1.   *Are men more likely than women to be married? Explain.*

The results of the cross-tabulation command may appear
confusing at first, but they are not truly mysterious. As indicated in the
first two lines of output, the rows of the table represent the various
marital status categories. The columns distinguish between females and
males. For instance, 135 women in the sample reported being divorced.

Our goal is to compare the responses of the men and the women.
However, simply looking at the frequencies could be misleading, since
the sample does not have equal numbers of men and women. With as
large a sample as we have here, it is more helpful to compare the
*percentage* of married men to the percentage of married women.

The cross-tabulation function can easily convert the frequencies to relative frequencies. We could return to the Cross-Tabulation dialog following the same menus as before, or take a slightly different path.

## Editing Your Most Recent Dialog

🖰 **Edit ➤ Edit Last Dialog**  This command will always return you to the most recent dialog box. Now we want the values in each cell to reflect frequencies relative to the number of women and men. To do so, check **Column Percents**, then click **OK**.

> 2. *Based on this new table, would you say that men or women are more likely to be married?*

🖰 Now try asking for **Row Percents** instead of column percents.

> 3. *Explain what the row percentages tell you.*

## More on Bar Charts

We can also use a bar chart to analyze the relationship between two variables. Let's first represent these data graphically. We will create a bar chart to display the same data that we just tabulated.

🖰 **Graph ➤ Bar Chart...**  The bar chart command was used in both prior sessions. In this case, since we want to display two variables (marital status and sex), we'll initially specify that we want a **Cluster** graph, as shown here.

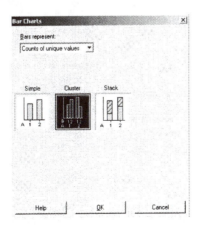

🖱 In the sub-dialog, select **MARITAL** as the Categorical variable, and then select **SEX** as the additional categorical variable, to produce the graph shown below.

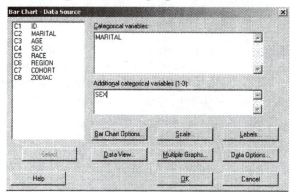

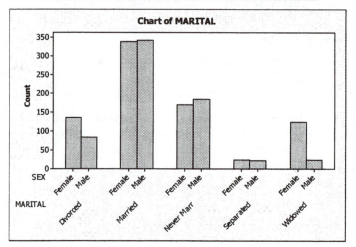

4.   *This graph displays the very same information as the first cross-tabulation. Which of the two displays represents the data more clearly, in your opinion? Explain.*

This graph displays raw counts (frequencies), rather than percentages of respondents. As such, it is still makes comparisons across samples or years difficult to interpret. A *stacked bar chart* is better suited to such comparisons.

🖱 **Graph ➤ Bar Chart...** In the initial dialog, select **Stack**.

🖰 In the sub-dialog, this time select **SEX** as the Categorical variable, and **MARITAL** as the additional categorical variable.

5.  *Describe what the green portion of each stacked bar represents.*

6.  *Repeat the stacked bar chart command, this time specifying* MARITAL *as the Categorical variable, and* SEX *as the additional one. How does this change the resulting graph?*

We can also use the **Bar Chart** command to analyze a quantitative variable. Suppose we wanted to compare the ages of the men and women in the survey. We might consider comparing the averages of the two groups.

🖰 **Graph ➤ Bar Chart...** In this initial dialog, use the **Bars represent** drop-down menu and select **A function of a variable**.

🖰 Next, select **Simple** and click **OK**.

🖰 Under **Function**, select **Mean**[2].

🖰 Choose **AGE** as the **Graph variable**, and **SEX** as the Categorical Variable. Click OK, and you will see this graph:

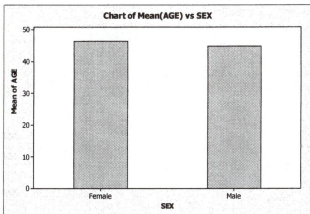

---

[2] As we will see in Session 4, the *mean* is the arithmetic average of a variable.

**7. *What does this graph show? What conclusion would you draw about the relative ages of men and women in the sample?***

**8. *Construct a similar bar chart comparing mean ages by marital status. Describe the noteworthy features of the graph.***

## *Comparing Two Distributions*

The graph of male and female ages compares the means of two distributions. How do the entire distributions compare? You already know how to create a histogram for a quantitative variable. Let's expand our knowledge of the histogram function, using the age variable. We begin by looking at the distribution of age across all respondents.

🖱 **Graph ➤ Histogram...** Select a **Simple** histogram, choose **AGE** as the variable, and click **OK**. You'll see the graph shown here.

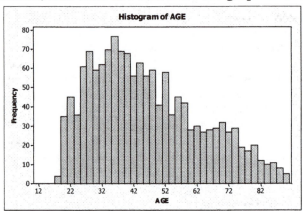

**9. *How would you describe the shape of this distribution?***

Let's see how the distribution of ages compares for male and female students. We'll create one histogram showing both distributions:

🖱 **Graph ➤ Histogram...** We need to indicate that the ages are to be displayed by gender, and that the graph should distinguish between the two. Therefore, select **With Outlines, Groups**.

🖱 As shown here, select **AGE** as the graph variable and **SEX** as the categorical variable.

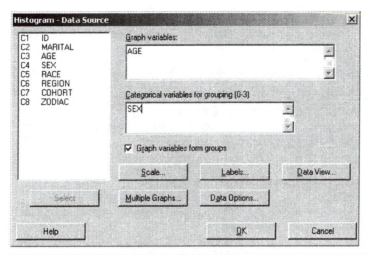

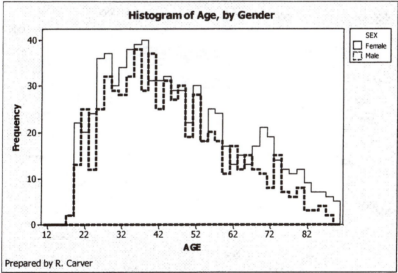

Prepared by R. Carver

On your screen the outline of the histogram for males will appear in red. Here, it is shown as a dashed line for clarity.

10. *Generally speaking, how do the center, shape, and spread of these two histograms compare?*

11. *Which gender has comparatively more young respondents in this sample?*

**12.** *Which gender has comparatively more respondents over the age of 60 in this sample?*

## Scatter plots to Detect Relationships

The prior example involved a quantitative and a qualitative variable. Sometimes, we might suspect a connection between two quantitative variables. In the student data, for example, we might think that taller students generally weigh more than shorter ones. We can create a *scatter plot* or XY graph to investigate.

🖰 Open the **Student** worksheet now.

🖰 **Graph ➤ Scatterplot...** Choose a **Simple** scatterplot.

🖰 In the main dialog box, select **Wt** as the Y (vertical axis) variable, and **Ht** as the X variable. Title your graph (use the **Labels** options). Click **OK**.

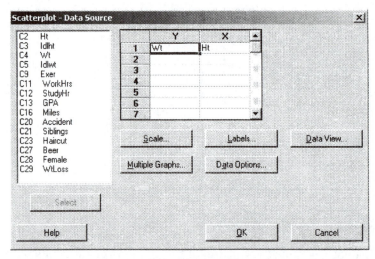

This will create the graph shown on the next page.

**13.** *What does the graph show you?*

**14.** *By eye, estimate the weight of a person who is 5'8" (or 68 inches) tall.*

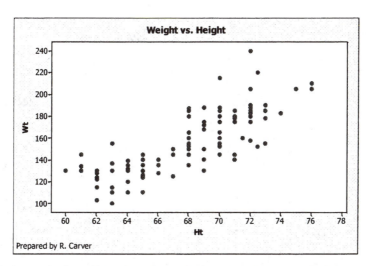

Prepared by R. Carver

It seems plausible that the relationship between height and weight might be different for male and female students. We can easily incorporate a *third* variable into this graph. We'll use two different approaches, and ask why we might select one approach over the other.

🖱 **Editor ➤ Panel...**[3]  Select **Gender** as the **By variable**.

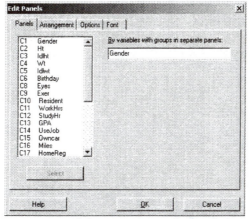

This creates one panel for women and one for men. As you look at these graphs, think about the *relationship* between height and weight.

---

[3] You can also find the Panel command by moving the cursor over the scatterplot, and right-clicking the mouse.

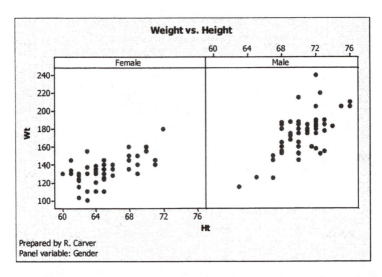

**15.    *What do these two graphs tells us about the relationship between height and weight for men and women?***

With these two graphs side by side sharing a vertical axis, it is easy to compare the weights of the two groups. However, it is a bit more challenging to compare the heights. Let's try another approach.

🖱 **Graph ➤ Scatterplot...** This time, select **With Groups**. Once again, **Wt** is on the Y-axis and **Ht** is on the X. In the **Categorical variables for grouping** box, select **Gender**.

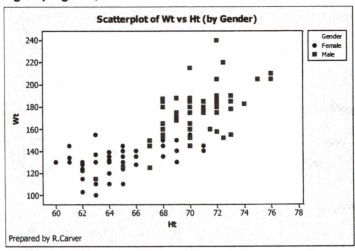

16.   *Compared to the two-panel scatterplot, what additional information (if any) does this graph convey?*

17.   *Can you make any generalizations about the heights and weights of men and women?*

18.   *Can you detect any outliers?*

## Time-Series Comparison of Two Variables

In the previous session, we discussed the difference between cross-sectional and time-series (or longitudinal) observations of a variable. Thus far, this session has focused on cross-sectional data. It is a simple matter to compare the movement of two sets of time series data. We will return to the United States demographic data that we met in the prior session.

🖱 **File ➤ Open Worksheet...** Open the file **US.MTW**.

🖱 **Graph ➤ Time Series Plot...** In the initial dialog, choose **Multiple.**

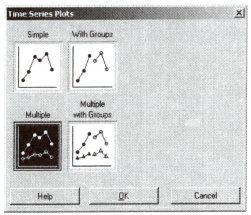

In this example, we'll look at the growth in United States federal revenues and expenditures during the period from 1965 to 2000. In our worksheet, **FedRecpt** represents total annual receipts, or revenue, from tax collections and other sources. **FedOut** is annual federal outlays, or expenditures. Both are measured in billions of dollars.

🖱 Complete the Time Series Plot dialog as shown here, and then click on **Time/Scale**.

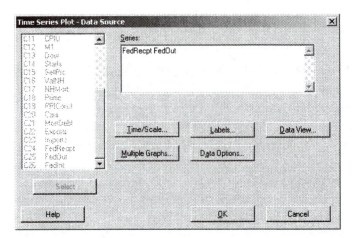

In the subdialog (see next page), choose **Stamp**, and select the variable Year as the *time stamp* for the horizontal axis. Click **OK** in both the sub- and main dialogs.

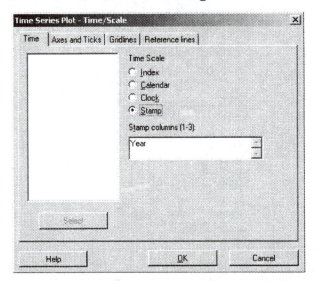

**19.   *Describe the relative growth in federal outlays and receipts; in what year does it appear that the US had its largest deficit (i.e., when were outlays most in excess of receipts)?***

## Moving On...

Create the tables and graphs described below. Refer to Appendix A for complete data descriptions. *Be sure to title each graph, including your name. Print these completed graphs.*

### Student

1. Generate a Histogram of the distribution of heights, separating men and women. Comment on the similarities and differences between the two groups.

2. Do the same for students' weights.

3. Create a scatter plot to analyze the relationship between students' actual weights (X) and the ideal weights (Y). In words, describe the relationship you observe.

4. Is the relationship between actual and ideal weight the same for male and female students?

5. Create a cross-tabulation of seatbelt usage and car ownership. Comment on any pattern you detect.

### GSSGeneral

6. Compare the mean ages of respondents across the categories of marital status. Offer an explanation for the differences you observe in average ages.

7. Does the distribution of marital status appear to vary by region of the United States? Explain.

8. Does the distribution of marital status appear to vary by race? Explain.

### Galileo

In the sixteenth century, Galileo conducted a series of famous experiments concerning gravity and projectiles. In one experiment, he released a ball to roll down a ramp. He then measured the total horizontal distance which the ball traveled until it came to a stop. The data from that experiment occupy the first two columns of the data file.

In a second experiment, a horizontal shelf was added to the base of the ramp, so that the ball rolled directly onto the shelf from the ramp.

Galileo recorded the vertical height and horizontal travel for this apparatus as well, which are in the third and fourth column of the file.[4]

9.  Construct a scatter plot for the first experiment, with release height on the X-axis and horizontal distance on the Y-axis. How would you describe the relationship between the two variables?

10. Do the same for the second experiment.

## Florida Votes

During the 2000 Presidential Election, there were a number of controversial occurrences in Florida. Among the controversies was the use of a "butterfly" ballot in Palm Beach County, which some voters claim to have found confusing. Specifically, the news media reported that voters intending to cast their votes for Al Gore might instead have selected Pat Buchanan.

11. First, let's look at the statewide totals for all candidates. Use the **Bar Chart** command for multiple Y's, choosing Function of a variable for the display option. Then, specify that the bars should represent the Sum of each column, and select all of the columns (C2-C11) *except* Total. Comment on the relative number of votes cast for President Bush, Vice President Gore, and Mr. Buchanan.

12. Use a simple scatter plot to graph the number of votes cast for Buchanan versus the number cast for Gore. Describe the overall pattern of voting behavior.

13. You may notice one point in the upper right of the graph. On the **Editor** menu, select **Brush**. Move the cursor to that point, and note the Row number that appears in the **Brushing** window on the left of your screen. In the data window, you can identify the county reporting that data point. What county is it, and what is unusual about the point?

---

[4] *Sources:* Drake, Stillman. *Galileo at Work*, (Chicago: University of Chicago Press, 1978); Dickey, David A. and Arnold, J. Tim "Teaching Statistics with Data of Historic Significance," *Journal of Statistics Education*, v.3, no. 1, 1995.

14. Create a similar scatter plot comparing votes cast for President Bush and those for Buchanan. Comment on what you see.

## US

15. Create a time-series plot that compares the growth in aggregate personal consumption and aggregate personal income; comment on what you see.

16. Create a scatter plot with personal income on the horizontal axis and consumption on the vertical. Discuss the relationship that appears in the graph.

17. Create a time-series plot that compares the growth in aggregate personal consumption and aggregate personal income; comment on what you see.

18. Create a time-series plot that compares the growth in the population of the United States and the number of cars in use in the U.S. Comment on what you see.

19. Now create a scatter plot with population on the horizontal axis and cars in use on the vertical. Discuss the relationship that appears in this graph, and compare it to what you saw in the prior graph.

## Mendel

Gregor Mendel's early work laid the foundations for modern genetics. In one series of experiments with several generations of pea plants, his theory predicted the relative frequency of four possible combinations of color and texture of peas.

20. Construct charts of both the actual experimental (observed) results and the predicted frequencies for the peas. Comment on the similarities and differences between what Mendel's theory predicted, and what his experiments showed.

## Texas Geog

This worksheet contains almanac data from the state of Texas.

21. What kind of relationship might you expect to find between a county's population in 1990 and its population in 2000? Construct a simple scatter plot with the 2000 Census counts on the vertical axis and the 1990 on the horizontal axis. Comment on the relationship that you observe.

22. What kind of relationship might you expect to find between the physical size of a county and its population? Construct a scatter plot of the 2000 population versus the land area (in square miles) of each county. Does there appear to be a relationship between geographical size and population?

23. What kind of relationship might you expect to find between the longitude and latitude of the county seat (the municipality housing county government) in all of the counties of Texas? Construct a scatter plot with Latitude on the Y-axis and Longitude on the X-axis. Why does this graph have this appearance? Would a graph of county locations in any U.S. state have an analogous appearance? Explain.

## Impeach

This file contains the results of the U.S. Senate votes in the impeachment trial of President Clinton.

24. The variable called **conserv** is a rating scale indicating how conservative a senator is (0 = very liberal, 100 = very conservative). Use a bar chart to compare the mean ratings of those who cast 0, 1, or 2 votes to convict the President. Comment on any pattern you see.

25. The variable called **Clint96** indicates the percentage of the popular vote cast for President Clinton in the senator's home state in the 1996 election. Use a bar chart to compare the mean percentages for those senators who cast 0, 1, or 2 votes to convict the President. Comment on any pattern you see.

26. Construct a cross tabulation of senators' party affiliation and their vote on the question of whether the President was guilty of perjury. Comment on what you find.

27. Construct a cross tabulation of senators' party affiliation and their vote on the question of whether the President was guilty of obstruction of justice. Comment on what you find.

## GSSSex1

28. Cross tabulates the responses to the questions about condom use and highest educational degree. Comment on what you observe.

29. Cross tabulate the responses to the questions about condom use and marital status. Comment on what you observe. How does your answer to this question affect your thinking about your previous answer?

## StateTrans

30. Use a scatter plot to explore the relationship between the number of fatal accidents in a state and the population of the state. Comment on the pattern, if any, in the scatterplot.

31. Use a scatter plot to explore the relationship between the number of fatal accidents in a state and the annual vehicle miles of travel within the state. Comment on the pattern, if any, in the scatterplot.

## Eximport

This worksheet contains time-series data about exports and imports in the U.S. economy. All amounts in the worksheet are in millions of dollars.

32. Create a multiple time-series plot for general imports and exports excluding military aid (**Exnoma**). Comment on the relative growth of imports and exports.

33. Create a scatter plot for the same two variables, and comment on their relationship, if any.

34. Create a multiple time-series plot for imports of foreign automobiles and imports of petroleum. Discuss noteworthy features of this graph.

# One-Variable Descriptive Statistics

## Objectives

In this session, you will learn to:

- Develop summary measures for a qualitative variable
- Compute measures of location and dispersion for a variable
- Detect symmetry or skewness in a distribution
- Create a box-and-whiskers plot for a single variable
- Compute z-scores for all values of a variable

## Computing One Summary Measure for a Variable

There are several measures of location (mean, median, mode, and percentiles), and of dispersion (range, variance, standard deviation, etc.) for a single variable. You can use Minitab to compute or generate these measures. We'll start with the mode of an *ordinal* variable.

🖱 **File ➤ Open Worksheet...** Select the file called **Student**. The file contains student responses to a first-day-of-class survey.

The variable called **Driver** holds students' responses to the question, "How would you rate yourself as a driver?"

🖱 **Stat ➤ Tables ➤ Tally Individual Variables...** Select the variable **Driver**, and click **OK**. Look at the results.

1.  *What was the modal response?*

2.  *What, if anything, strikes you about this frequency distribution?*

**3.** *How many students are in the "middle?"*

**4.** *How do these students' define "average?"*

🖱 Before continuing, **Edit your last dialog**, and check **Percents, Cumulative Counts**, and **Cumulative Percents**. You will see this:

```
Tally for Discrete Variables: Driver

        Driver  Count  CumCnt   Percent  CumPct
Above Average      63      63     56.76   56.76
      Average      46     109     41.44   98.20
Below Average       2     111      1.80  100.00
          N=      111
```

**5.** *What does each of the columns in the Tally tell you?*

**Driver** is a qualitative variable with three possible values. Some categorical variables have only two values, and are known as *binary* variables. Gender, for instance, is binary. In this dataset, there are two variables representing a student's sex. The first, which you have seen in earlier sessions, is called **Gender**, and is a Minitab text variable, assuming values of "Female" and "Male." The second is called **Female**, and is a numeric variable equal to 0 for men and 1 for women. If we wanted to know the *proportion* of women in the sample, we could tally either of the worksheet variables. Alternatively, we could compute the *mean* of **Female**.

🖱 **Calc ➤ Column Statistics...** In this dialog box, choose the radio button for **Mean**, and select the variable **Female** as the input variable. Click **OK**.

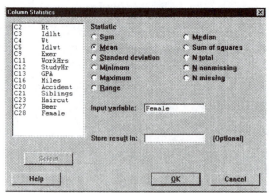

**6.** *What percent of the respondents in this survey are women? Compare this to the results if you Tally the variable called Gender, and comment on what you find.*

Now let's move on to a *quantitative* variable: the number of brothers and sisters the student has. The variable is called **Sibling**.

🖱 Use the **Tally individual Values** command to find the *mode* of the variable, **Sibling**. (You should de-select **Cumulative Counts** and **Cumulative Percents** in this dialog.)

Now let's find the other measures of central tendency. There are two ways to do this. If we simply want to know the mean or the median, we can compute it alone with the Column Statistics command.

🖱 **Calc ➤ Column Statistics...** As you just did for **Female**, find the mean for **Sibling**.

Look in the Session Window, and you will see the Mean number of siblings per student.

**7.** *Do you think that anyone in the class had this number of siblings? If not, then in what sense does the mean describe the variable?*

**8.** *If you needed to describe the students' responses to this question, would you prefer to tally these results or compute the mean or median? Explain.*

## Computing Several Summary Measures

Suppose you wanted to know both the mean and standard deviation, or several other descriptive measures. You could proceed as we did above, but the Column Statistics command will only compute one statistic at a time. To get several summary measures at one time, we use a different command.

🖱 **Stat ➤ Basic Statistics ➤ Display Descriptive Statistics...** Select the variables **Ht** (height in inches) and **Wt** (weight in pounds). You will see the output shown on the next page. The default output provides ten different descriptive statistics for each of the variables.

```
Descriptive Statistics: Ht, Wt

Variable  N    N*    Mean Median    StDev  SE Mean  Minimum  Maximum        Q1
Ht       111    0  68.068     68   3.7504  0.35598       60       76     65.00
Wt       110    1 156.118    153  28.2942  2.69775      100      240    134.75

Variable Q3
Ht          71
Wt         180
```

Specifically, the command provides these summary measures:

| N | Number of observations in the sample for this variable |
|---|---|
| N* | Number of missing observations for this variable |
| Mean | The sample mean, or $\bar{x} = \dfrac{\sum x}{n}$ |
| Median | The sample median (50th percentile) |
| StdDev | The sample standard deviation, or $s = \sqrt{\dfrac{\sum (x - \bar{x})^2}{n-1}}$ |
| SEMean | The standard error of the mean. This measure becomes important later in your course, and its interpretation should be held off until then. The formula for the standard error is $s_{\bar{x}} = \dfrac{s}{\sqrt{n}}$ |
| Min | The minimum observed value for the variable |
| Max | The maximum observed value for the variable |
| Q1 | The first quartile (25th percentile) for the variable |
| Q3 | The third quartile (75th percentile) for the variable |

Compare the mean and median for the two variables. Doing so can shed light on the *shape* of the distribution, revealing whether it is symmetric or skewed. In a single-peaked distribution, if the mean exceeds the median, we say that a distribution is *positively*, or *right*, *skewed*. If the mean is less than the median, we say that a distribution is *negatively*, or *left*, *skewed*.

9.  **Are these two variables symmetric or skewed?**

10. **Does either of the two appear to have some outliers skewing the distribution? Look at a dotplot of the two variables to check your thinking; does the plot confirm your conclusion?**

The Stat menu offers an option that relates the summary statistics to the graphs you worked with in earlier sessions. For example, let's take a closer look at the heights.

🖱 **Stat ➤ Basic Statistics ➤ Graphical Summary...** Now select only **Ht.**

In the graphical summary (see below), you get all of the summary measures as before along with some new ones, and you also see a histogram with a symmetrical bell-shaped curved superimposed, a *box-and-whiskers* plot, and two *confidence interval* graphs. The additional information refers to topics that we will discuss in later sessions. At this point in the course, though, you know what most of this output means.

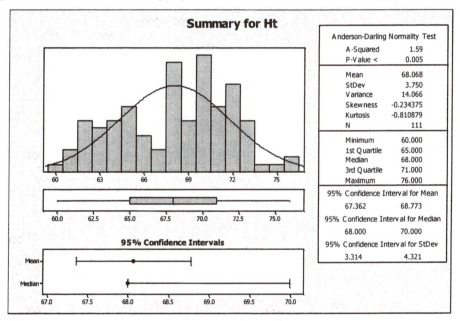

The *skewness* statistic is a numerical measure of symmetry. At this stage of your statistics education, it is helpful to understand that the sign of the statistic indicates the direction of skewness, and that the absolute value of the statistic tells us the extent of skewness. In comparing two distributions, the one with the smaller skewness statistic is more symmetrical.

The *kurtosis* statistic measures the steepness of a single-peaked distribution. It helps us gauge how close the distribution is to a *normal distribution*, which is introduced in Session 8.

## The Empirical Rule

Among the descriptive statistics is the *standard deviation*, which is a measure of dispersion or spread. Unlike the mean and median, which point to a location on the number line, the standard deviation requires more complex interpretation. Ordinarily, we will speak of percentages of a data set that are concentrated within a given number of standard deviations from the mean.

For distributions that are generally mound-shaped (single-peaked, without severe skewness), we can use the *Empirical Rule* to interpret the standard deviation. The Empirical Rule is a practical guideline that relates the mean and standard deviation as follows:

- About two-thirds (68%) of all observations lie within one standard deviation of the mean.
- Roughly 95% of all observations lie within two standard deviations of the mean.
- Nearly all (99.7%) observations lie within three standard deviations of the mean.

The last provision implies that, for reasonably symmetrical data sets, the *range* will be no more than six standard deviations.

Let's see how well the Empirical Rule applies to the height data, despite the fact that this distribution has two peaks. The mean and median are similar, suggesting a relatively symmetric distribution. The standard deviation is 3.75 inches, and the range equals the 76–60, or 16 inches. Notice that the mean is just about halfway between the minimum and maximum values. We can also quickly calculate that the range is 16/3.75, or 4.27 standard deviations. Thus, all of the students are certainly within three standard deviations of the mean.

The sample mean minus one standard deviation is roughly 68.07–3.75 or 64.32 inches; the sample mean plus one standard deviation is 71.82 inches. According to the rule, approximately two-thirds of the 111 students (about 74 students) should be between 64.3 and 71.8 inches tall. A stem-and-leaf display of heights indicates that 65 students (about 59%) fall in this interval. If we add in the students who are 64 inches tall, we have 72 students in the interval.

Similarly, the Empirical Rule predicts that 95% of the students' heights are within two standard deviations of the mean. In other words, all but five or six students should be in the height interval.

11.    *Find those two heights (in inches) that lie two standard deviations above and below the mean.*

**12.  *Use a stem-and-leaf display to discover how many students are either shorter or taller than the interval you just found.***

The Empirical Rule provides very useful approximations to help us characterize the distribution of a data set. In this example, the predictions were not perfect, but were reasonable approximations, even with a distribution that is not ideally mound-shaped.

## Box-and-Whiskers Plots

The descriptive statistics command generates the ***five-number summary*** for a variable (minimum, maximum, first and third quartiles, and median). A Box-and-Whiskers Plot (or a *boxplot*) visually displays the five-number summary. Boxplots are particularly helpful in comparing two distributions.

🖱 **Graph ➤ Boxplot...**  Choose **One Y With Groups**, as shown.

🖱 Select **Ht** as the graph variable, and **Gender** as the categorical variable, and click **OK**.

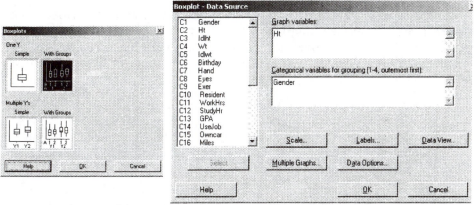

In the resulting graph window, you'll see five-number summaries of the heights of men and women in the class. Consider the females. We have a box in the center of the figure with a line across its middle. The top and bottom of the box represent the third and first quartiles, respectively, and the line shows the median. The "whiskers" extend either to the minimum and maximum values, or to one-and-one-half of the interquartile range. Asterisks, if any, indicate outlier values.

13. *What does this graph suggest about the center and spread of the variable for the two sub-groups?*

14. *Where are the median and the quartiles in the graph?*

 Try another boxplot for **Wt**, also using **Gender** as the grouping variable.

15. *How do the two boxplots for weight compare?*

16. *How do the weight and height boxplots compare to one another?*

17. *How do you explain the differences?*

## Standardizing a Variable

 Now open the file **Marathon.** This file contains the finish times for all wheelchair racers in the 100th Boston Marathon.

18. *Find the mean and median finish times. What do these two statistics suggest about the symmetry of the data?*

Assuming that few of us are experts on the subject of wheelchair marathons, it may be difficult to know if a particular finish time is "good" or not. It is sometimes useful to *standardize* a variable, so as to express each value as a number of standard deviations above or below the mean. Such values are also known as *z-scores*.

 **Calc ➤ Standardize...** This command computes a z-score for each value of the Input variable and stores the computed values in a specified worksheet column. In this case, the input variable is **Minutes**. Move your cursor into the box marked **Store results in** and type **zscore**, and click OK.

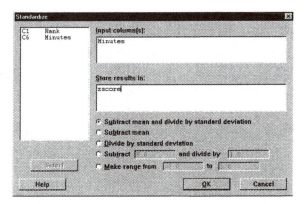

Now look at C7 in the Data Window. Since the racers are listed by finish rank, the first z-score value belongs to the winner of the race, whose finishing time was well below average. That's why his z-score is negative, indicating that his time was less than the mean.

Compare the z-scores of the top two racers.

**19.** *How does the difference between them compare to the difference between finishers #2 and #3?*

**20.** *How does the difference between the top two racers compare to the difference between the last two finishers? What does this suggest about the difference between a ratio variable (like* Minutes*) and an ordinal variable (like* Rank*)?*

**21.** *Did anyone finish with a z-score of approximately 0? What does that indicate?*

## Moving On...

Now use the commands illustrated in this session to answer these questions. Where appropriate, indicate which statistics you computed, and why you chose to rely on them to draw a conclusion.

### Student

1. What was the mean amount paid for a haircut?

2. What was the median amount paid for a haircut?

3. Which measure of central location seems to more accurately reflect the amounts that students paid for haircuts? Explain your thinking.

4. What does the comparison of the mean and median suggest about the shape of the distribution of haircut costs?

5. Create a box-and-whiskers plot for haircut prices; in what way does this plot confirm your answer to the previous question?

6. 25% of the students paid more than how much for a haircut?

## Output

This file contains data about industrial production in the U.S. from 1945–1996. Column 6 represents the degree to which the productive capacity of all U.S. industries was being utilized. Column 7 has a comparable figure, just for manufacturers.

7. During the period in question, what was the mean utilization rate for all industrial production? What was the median? Describe the symmetry and shape of the distribution for this variable. (Hint: Use a graphical summary.)

8. During the period in question, what was the mean utilization rate for manufacturing? What was the median? Describe the symmetry and shape of the distribution for this variable.

9. Construct a simple box-and-whiskers plot for manufacturing utilization; describe the symmetry and shape of the distribution.

10. Use the Empirical Rule to find the approximate range of this distribution; compare your results to the actual range. How well does the Empirical Rule work in this case?

11. Compare the standard deviations of overall utilization or manufacturing utilization; which varied more?

12. Comment on similarities and differences between the two variables.

## Sleep

This file contains data about the sleeping patterns of different animal species.

13. Construct simple box-and-whiskers plots for **Lifespan** and **Sleep**. For each plot, explain what the "landmarks" on the plot tell you about each variable.

14. The mean and median for the **Sleep** variable are nearly the same (approximately 10.5 hours). How do the mean and median of **Lifespan** compare to each other? What accounts for the comparison?

15. According to the dataset, "Man" (row 34) has a life span of 100 years, and sleeps 8 hours per day. Determine, in terms of quartiles, where humans fall among the species for each of the two variables.

16. Sleep hours are divided into two types: "dreaming" and "non-dreaming" sleep. On average, do species spend more hours in dreaming sleep or non-dreaming sleep?

## Water

These data represent water usage in 221 regional water districts in the United States for 1985 and 1990.

17. Column 17 (**to-cufr85**) is the total amount of fresh water used for consumption (drinking) in 1985. On average, how much drinking water did regions consume in 1985?

18. Column 34 (**percentcu**) is the percentage of all *fresh* water devoted to consumptive use (as opposed to irrigation, etc.) in 1985. What percentage of fresh water was consumed, on average, in water regions during 1985?

19. Which of the two distributions was more heavily skewed? Why was that variable less symmetric than the other?

## Labor1

This file contains monthly data about the U.S. labor force from January 1948 through March 1996.

20. According to the Empirical Rule, labor force participation among teens was between $a$ and $b$ for approximately 95% of the months recorded. Find the appropriate values for $a$ and $b$, and explore the data to see if the rule accurately describes the data set. Comment on what you find.

21. Compute the descriptive statistics for labor force participation rates among males, females and teens. Comment on the similarities and differences in the measures of center and spread for these three variables.

22. Now construct a time-series plot for each of the three variables. What important pattern appears in the plot? Do the descriptive statistics capture this pattern? What does this illustrate about the choices we might make in describing this set of data?

## Triathlon

This file contains the results of the Women's Triathlon during the 2000 Olympic games. Finishing times are shown both as hours:minutes:seconds:hundreths and in minutes.

23. Compute the descriptive statistics for **Minutes** and use the maximum and minimum to compute the range of the variable. Is the range more or less than six standard deviations?

24. Standardize the variable called minutes; how many standard deviations separated the two fastest athletes? How many standard deviations separated the two slowest athletes? Do you think many athletic competitions turn out this way? Explain.

## JFKLAX

This file contains flight delay data for all commercial flights between JFK (New York) and LAX (Los Angeles) airports for one randomly selected day.

25. How long was the average departure delay that day? Hint: recall that both the mean and the median are considered to be averages.

26. How long was the average arrival delay that day?

27. Which varied more: departure delays or arrival delays?

28. Which statistical technique would you use to determine the percentage of flights that were late in arriving? Use the technique to estimate or determine the percentage of late arrivals.

# Two-Variable Descriptive Statistics

## *Objectives*

In this session, you will learn to:

- Compute the coefficient of variation
- Compute measures of location and dispersion for subsamples of a variable
- Compute the covariance and correlation coefficient for two quantitative variables

## *Comparing Dispersion with the Coefficient of Variation*

In Session 4, you learned to compute descriptive measures for a single variable, and to compare these measures for different variables. Often, the more interesting and important statistical questions call upon us to compare two sets of data, or to explore possible relationships between two variables. This session deals with techniques for making such comparisons and describing such relationships for quantitative data.

Comparing the means or medians of two variables is straightforward. On the other hand, when we compare the dispersion of two variables, it is sometimes helpful to take into account the magnitude of the individual data values. For instance, suppose we sampled the heights of mature maple trees and corn stalks. We could anticipate the standard deviation for the trees to be larger than that of the stalks, simply because the heights themselves are so much larger. What we need is a *relative measure* of dispersion. That is what the Coefficient of Variation (CV) is.

The CV is the standard deviation expressed as a percentage of the mean. Algebraically, it is:

$$CV = 100 \cdot \left( \frac{s}{\bar{x}} \right)$$

In Minitab, we can perform calculations like the CV and store the results either in columns or in *constants*. Calculations that result in new *variables*, (i.e. generate a value for each observation of one or more existing variables) are stored in columns; calculations that yield a single result are stored in constants. The CV formula will generate a constant.

We'll use a single dataset through much of this session. The examples deal with traffic and highway travel in the United States. We'll begin by looking at the roads, specifically comparing the lengths of urban and rural roads and highways throughout the 50 states and the District of Columbia.

🖱 Open the file called **StateTrans**.

🖱 **Stat ➤ Basic Statistics ➤ Display Descriptive Statistics...** Select the variables **RLM** and **ULM**. These refer to Rural Lane Miles and Urban Lane Miles. A *lane mile* is a measure of roadway capacity; a four-lane highway 25 miles in length represents 4 x 25, or 100 lane miles.

1.  *Which variable had the higher mean? How does this result square with your "mental map" of the U.S.?*

2.  *Notice the standard deviations. Which set of measurements varied more? Discuss why one variable would vary more than the other.*

The comparison is sharpened if we look at the Coefficient of Variation.

🖱 **Calc ➤ Calculator...** Minitab represents constants as K1, K2, etc. In the box marked **Store result in variable**: type **K1**. Move your cursor into the **Expression** block.

🖱 In the Expression block, type the following (be careful to copy it precisely; see the completed dialog on the next page).

```
100*stdev('RLM')/mean('RLM')
```

NOTE:  When typing the apostrophes (single quotations marks), use the right-hand key with single and double quotation marks. You can also insert the expression 'RLM' by double-clicking on it in the variable list.

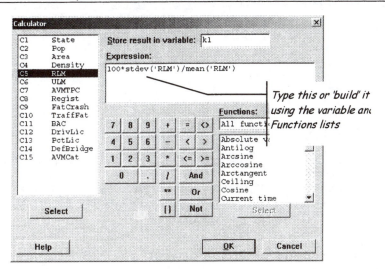

☝ Click **OK**. You will not see any obvious result of your computation yet, but if you see an error message, try again.

☝ Now repeat the computation for urban lane miles, storing the result in constant **K2**. K1 and K2 now contain the CV's for rural and urban lane miles, respectively.

☝ Now choose **Window ➤ Project Manager**[1]  In the tree diagram to the left, select the **Constants** folder, and you should see the window shown on the next page.

3. *Compare the two coefficients of variation with the two standard deviations you computed earlier. What do you notice about the degree of variation for these two variables?*

4. *So, which varies more: RLM or ULM?*

---

[1] You could also choose **Data ➤ Display** data... and select **K1** and **K2**.

5. **What characteristics of the fifty states might explain the difference you find in the extent of variation?**

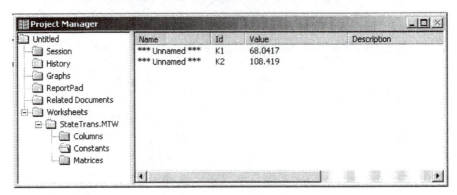

6. **The standard deviations for the two variables were expressed in number of miles. What are the units for the coefficient of variation?**

## Descriptive Measures for Subsamples

The transportation infrastructure of the United States varies in many ways from state to state. We've just seen that states vary widely in roadway capacity. There is also wide variation in how much residents drive in a year. We have two variables that represent the amount of driving in each state—one quantitative, the other qualitative. The quantitative variable, **AVMTPC**, is the Average Vehicle Miles of Travel per Capita in 1998. In other words, it is the average distance traveled per person in the state. The second variable, **AVMCat**, is a binary categorical variable that simply classifies the states as having high or low annual-vehicle-miles.[2]

We might wonder if states whose citizens drive a lot have more miles of roadway than the rest. We can compute separate descriptive measures for these two groups of states. To do so, we invoke the "By variable" option in the descriptive statistics command:

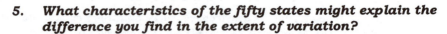 **Stat ➤ Basic Statistics ➤ Display Descriptive Statistics...** Select both **RLM** and **ULM** as the variables, and move your cursor into the block marked **By Variable (optional)**, and select **AMVCat**.

---

[2] Low AVM states are those below the median **AVMTPC**; high states are above the median.

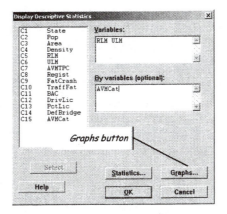

🖱 Before clicking **OK**, click on the **Graphs** button. In the **Graphs** dialog, select **Boxplot of Data.**

Two graph windows will open, one of which is shown below.

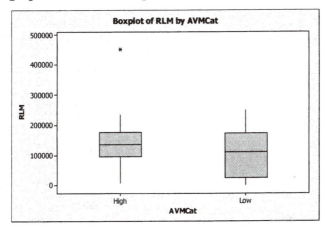

**7.** ***What do the two boxplots indicate about the comparative center and dispersion of rural and urban lane miles for high- and low-travel states?***

Find the numerical results in the Session Window. These should look familiar, with one new twist. For **RLM**, two lines of output appear. The first refers to those observations with high annual vehicle miles, and the second refers to the low-mileage subsample. Take a moment to reconcile these statistics with the two boxplots.

> **8.** *Look in the worksheet to find your home state, taking note of its RLM and ULM values. Where is your state positioned in the two boxplots?*

## *Measures of Association: Covariance and Correlation*

We have just analyzed a relationship between a quantitative variable (lane miles) and a qualitative variable (high vs. low travel). Sometimes we may be interested in a possible relationship or association between two *quantitative* variables. For instance, in this dataset we might expect that there is a relationship between the number of drivers licenses issued in the state (**DrivLic**) and the population of the state (**Pop**). Specifically, the population data is from the 2000 Census, and the license data, in thousands, is from 1998.

🖱 **Stat ➤ Basic Statistics ➤ Display Descriptive Statistics...** Clear **AVMCat** from the **By variable** option. Select the variables **DrivLic** and **Pop**.

> **9.** *Do the boxplots and statistics make sense to you? Remember that* DrivLic *is expressed in thousands.*

🖱 **Graph ➤ Scatterplot...** Create a simple scatter plot with **DrivLic** on the Y-axis, and **Pop** on the X-axis.

> **10.** *Do you see evidence of a relationship? Describe what you see.*

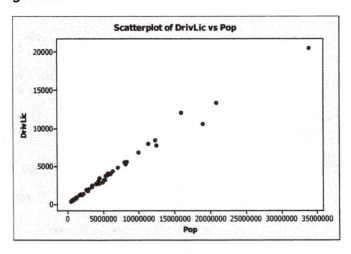

A graph like this one shows a strong tendency for X and Y to "co-vary." It is clear that large states have many licensed drivers.

There are two common statistical measures of co-variation. They are the *covariance* and the *coefficient of correlation*. In both cases, they are computed using all available observations for a pair of variables. The formula for the sample covariance of two variables, $x$ and $y$, is:

$$\text{cov}_{xy} = \frac{\sum (x_i - \bar{x})(y_i - \bar{y})}{n-1}$$

The sample correlation coefficient[3] is:

$$r = \frac{\text{cov}_{xy}}{s_x s_y}$$

where:

$s_x$, $s_y$ are the sample standard deviations of $x$ and $y$, respectively.

Though these formulae may look daunting, it is easy to compute each of them.

**Stat ➤ Basic Statistics ➤ Covariance...** Select the variables **DrivLic** and **Pop**, and click **OK**. You will see the following in the Session Window.

```
                   DrivLic                Pop
DrivLic          14921227
Pop           23671058829 3.79952E+13
```

In this triangular arrangement of values (known as a Variance-Covariance Matrix), the Covariance of **DrivLic** and **Pop** appears at the intersection of the respective column and row. The covariance of is 14,921,227. What, then, are the other two numbers?  They are the *sample variances* ($s^2$) of the two variables. You can verify this by consulting the descriptive statistics for both.[4]

A correlation coefficient (symbolized $r$) always takes a value between $-1$ and $+1$. Absolute values near 1 are considered "strong"

---

[3] Formally, this is the Pearson Product Moment Correlation Coefficient, known by the symbol, $r$.

[4] The sample variance of **Pop**, expressed as 3.78852E+13 is expressed in scientific notation: $3.79 \times 10^{13}$.

correlations; that is, the two variables have a strong tendency to vary together. Absolute values near 0 are weak correlations, indicating very little relationship or association between the two variables.

🖱 **Stat ➤ Basic Statistics ➤ Correlation...** Again, select the variables **DrivLic** and **Pop**, and click **OK**. You will find the correlation coefficient in the Session Window.

> *11.* **What is the correlation between these two variables? Would you say that it is a strong or weak correlation?**

Variables can have strong sample correlations for a number of possible reasons. It may be that one is the cause of the other, that a third variable causes both of them, or that their observed association in this particular sample is merely a coincidence. As you will learn later, correlation is an important tool in statistical reasoning, but we must never assume that correlation implies causation.

## Moving On...

Use the commands and techniques presented in this session to answer the following questions. Explain your choice of statistics in responding to each question.

### Impeach

This file contains data about the U.S. senators who voted in the impeachment trial of President Clinton.

1. Compare the mean home-state vote percentage for Clinton in the 1996 election among Democratic and Republican senators, and comment on what you find.

2. what is the correlation between the number of votes a senator cast against President Clinton in the trial and the number of years left in the senator's term? Comment on the strength of the correlation.

### GSSGeneral

This worksheet contains data from the 1998 General Social Survey.

3. Compare the mean and median age for men and women. Describe what you find.

4. Compare the standard deviation and the coefficient of variation for men and women. Comment on the degree of variation among the ages of men and women in this sample.

## GSSHousehold

This dataset has more data from the 1998 General Social Survey, all related to household composition.

5. The variable called **HOMPOP** represents the number of people living in the respondent's household. Before looking at the data, try to predict how the values of this variable might compare between men and women.

6. Compute relevant descriptive measures for **HOMPOP**, comparing male and female respondents. Comment on what you find.

7. Find the correlation between the number of people living in the household and the number of generations living in the household. Is the correlation strong or weak? What might account for this degree of correlation?

## Bodyfat

This dataset contains body measurements of 252 males.

8. What is the sample correlation coefficient between neck and chest circumference? Suggest some reasons underlying the strength of this correlation.

9. What is the sample correlation coefficient between biceps and forearm? Suggest some reasons underlying the strength of this correlation.

10. Which of the following variables is most highly correlated with body fat percentage (**FatPerc**): Age, Weight, Abdomen circumference, or Thigh circumference?

## Michelson

In 1879, A.A. Michelson took a series of measurements of the velocity of light in air over a distance of 600 meters, refining techniques earlier developed by Foucault. In 1882, he used a modified methodology, and recorded additional readings. This worksheet contains some of Michelson's results. Column C1 contains the measurements of light speed, recorded in kilometers per second (km/sec). Column C2 contains the year the measurements were collected. We want to compare the 1879 and 1882 results.

11. Which measuring technique gave Michelson more consistent results (hint: look at the standard deviations).

12. Look at the means and medians of the two sets of measurements; would you say these distributions are symmetrical or asymmetrical? Make appropriate graphs to check your reasoning.

13. Compare the 1879 and 1882 mean measurements; did Michelson's estimate of the speed of light differ substantially in these two sets of data?

## Sleep

This worksheet contains data about the sleep patterns of various mammal species. Refer back to Session 4 for more information.

14. Using appropriate descriptive and graphical techniques, how would you characterize the relationship (if any) between the amount of sleep a species requires and the mean weight of the species?

15. Using appropriate descriptive and graphical techniques, how would you characterize the relationship (if any) between the amount of sleep a species requires and the life span of the species?

## Water

In Session 4, you investigated column 17, representing the total fresh water consumption in 1985. Column 33 contains comparable data for 1990.

16. Compare the means and medians for these columns, as well as boxplots. Did regions consume more or less water, on average, in 1990 than they did in 1985? What might explain the differences five years later?

17. Compare the coefficient of variation for each of the two variables. In which year were the regions more varied in their consumption patterns?

18. Construct a scatter plot of Fresh Water consumption in 1990 (Column 33) versus the regional populations in that year (Column 18). Also, compute the correlation coefficient for the two variables. Is there evidence of a relationship between the two? Explain.

## MFT

These are the scores on a Major Field Test (MFT) for some college seniors. The MFT measures students' achievement in their major field of study.

19. The columns labeled **Sub1** through **Sub4** are student scores on the four subsections of the exam. Possible scores on each subsection range from 20 to 80. On which part of the exam did these students do best? Explain your thinking.

20. On which subsection of the exams did these students have the most consistent scores?

21. Compute the correlation between students' Total scores on the Major Field Test and their GPAs. Comment on the strength of the relationship, and suggest some plausible reasons for the observed strength.

22. Similarly, compute the correlation between MFT Total score and Math SAT score; do the same for Verbal SAT score. Explain what the correlations tell you.

# Session 6

## Elementary Probability

### Objectives

In this session, you will learn to:
- Simulate random sampling from a population
- Draw a simple random sample from a set of observations
- Recode worksheet data for analysis

### Simulation

Thus far, all of our work with Minitab has relied on observed sets of data. Sometimes we will want to exploit the program's ability to *simulate* data which conforms to our own specifications.[1] In the case of experiments in classical probability, for instance, we can have Minitab simulate flipping a coin 10,000 times, or rolling a die 500 times.

### A Classical Example

Imagine a game spinner with four equal quadrants such as the one illustrated here. Suppose you were to record the results of 1000 spins. *What do you expect the results to be?*

We can simulate 1000 spins of the spinner by having Minitab calculate some random data:

---

[1] More properly, these should be called *pseudo-random* samples, since the computer program follows an algorithm to generate them. In that sense, they are not truly random.

🖰 **Calc ➤ Random Data ➤ Integer...** This command creates random integers to our specification. We must specify a range of allowable values, a number of observations, and a column location for the new data. Complete the dialog as shown here:

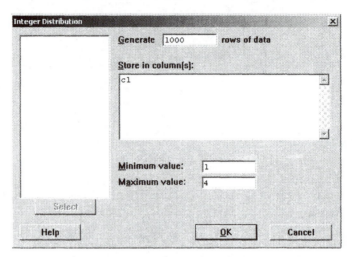

This will create a new variable, representing a random sample of 1000 spins. In the first column of the worksheet, you will see a list of values ranging from 1 through 4, simulating 1000 spins of the wheel.

> 💻 NOTE: Because these are random data, your data window will be unique. In other words, if you are working alongside a partner on another computer, her results will differ from yours.

1.  *If you were to tally the values in C1, how many 1s would you expect to find? Why?*

🖰 **Stat ➤ Tables ➤ Tally Individual Variables...** Tally the data in **C1**, requesting counts and percents.

2.  *What should the relative frequency (i.e. percent) be for each value? Do all of your results match the theoretical value exactly? To the extent that they differ, why don't they match exactly?*

Recall that classical probabilities give us the long-run relative frequency of a value. Clearly, an experiment with only 1000 spins does not adequately represent the "long-run," but this simulation may help you to understand what it means to say that the probability of spinning a "3" (or any other value) equals 0.25. In the standard language of probability, if we consider spinning a 3 to be an *event* (let's call it *A*), we would write *P(A)*=.25.

## *Observed Relative Frequency as Probability*

As you know, many random events do not result from classical probability experiments, and we must rely on observed relative frequency. In this part of the session, we will direct our attention to some of our General Social Survey (GSS) data. The GSS asked "Do you believe in Heaven?" These respondents gave five different answers: *definitely not, probably not, probably yes, definitely yes,* and *don't know*.

**3.**  ***How is this different from a five-quadrant spinner? Will all five answers occur with equal likelihood?***

🖱 **File ➤ Open Worksheet...** Open the **GSSRelig** worksheet.

🖱 **Stat ➤ Tables ➤ Tally Individual Variables...** Tally the variable **Heaven**, selecting Counts and Percents.

**4.**  ***What do these relative frequencies indicate about U.S. adults' belief in Heaven?***

```
Tally for Discrete Variables: HEAVEN

         HEAVEN  Count  Percent
     Don't know     71     5.80
No definitely not     75     6.13
 No probably not     90     7.35
  Yes definitely    766    62.58
    Yes probably    222    18.14
             N=   1224
            *=    221
```

**5.**  ***If you were to choose one person at random from this group, what is the probability that you select a person who believes that Heaven definitely exists?***

**6.**  ***What's the probability of choosing a person who believes in Heaven to some extent?***

In question 5, we asked about the probability of a single event, *P(believes definitely)*, which we might write symbolically as *P(BD)*. In question 6, we asked about a *union* of two events: *P(believes definitely* OR *believes probably*), which we might write as *P*(*BD* or *BP*), or alternatively as *P*(*BD* U *BP*).

Let's randomly select just one person and see what happens.

🖰 **Calc ➤ Random Data ➤ Sample from Columns...** We will sample one row from **HEAVEN**, and store the result in an empty column, which we'll name **Sample**. Complete the dialog as shown:

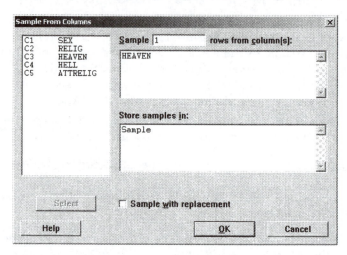

Look in the first row of the last column in the data window. There will be a phrase in the cell.

**7.   *Which phrase is it? Do you think your classmate at the next computer is looking at the same phrase?***

From the tally we know that almost 63% of those responding hold a definite belief in Heaven. That notwithstanding, knowing the relative frequency isn't much help in predicting any one person's response. On the other hand, it does help us predict what *pattern* we'd see if we asked 50 people. Now let's repeat this experiment 50 times:

🖰 Edit your last dialog and request 50 rows of data.

🖰 Tally the values in **Sample**.

**8.   *Do the relative frequencies of* Sample *match those of* Heaven*? How do your two tallies compare?***

## Recoding Alphanumeric Data

In the prior example, the variable of interest had five different reported values. Suppose we really wanted to divide respondents into two groups: those who report a belief in Heaven and those who do not. In other words, we'll lump the two positive response categories together, forming their *union*. We can create a new variable that combines the positive responses into one category, and all other answers into another.

9. **Look back at your tally. If you choose one person randomly, what is the probability that the person believes that Heaven exists?**

Now let's create our new variable, and *recode* the qualitative data to suit our purposes more directly.

🖰 **Data ➤ Code ➤ Text to Text…** We want to make a new variable (call it **HEAVEN2**) based on the data in **HEAVEN**, leaving the original variable unchanged.

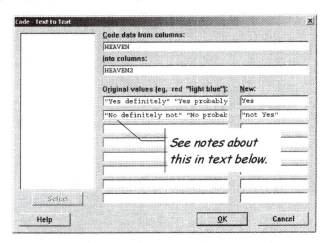

Complete the dialog as shown here, taking care to type the original data categories exactly as they appear in your original tally. In particular, pay attention to capitalization. Also note that you must type the phrases in double quotation marks (e.g. "Yes definitely") and allow a space between the phrases. Both sets of original values will extend beyond the text box; just keep typing and the line will scroll horizontally.

If you look in the Data Window, you'll find the new variable **HEAVEN2**. As you scroll through the column, you should find that each cell either says "Yes", "not Yes", or nothing at all (representing people who did not respond to the original question.

> **10.** *Now tally the responses in* HEAVEN2. *What is the probability of randomly selecting a person who answered "Yes"? How does this compare with your answer to question 9 above?*

## Joint and Conditional Probability

Thus far, we have been focusing our investigations on a single random experiment. We have selected a person and noted if the person believes in Heaven. Now let's take a further step by considering the interaction of two events. Specifically, let's see how men and women view the question.

In the language of elementary probability, let's define two events. Event "H" occurs if a randomly chosen person indicated a belief in Heaven. Even "F" occurs if a randomly chosen person is female. We want to estimate both the simple (*marginal*) probabilities of H and F, as well as their *joint* probability. The joint probability is the probability that they occur together in the same person. This is also known as the probability of the *intersection* of the two events, symbolized as $P(H \cap F)$.

**Stat ➤ Tables ➤ Cross tabulation and Chi-Square...** As shown in the dialog on the next page, the variables to select are **HEAVEN2** and **SEX**. Request **Counts** and **Total percents** to obtain frequencies and relative frequencies.

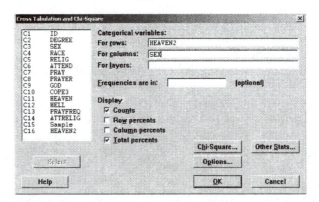

Look at the table in the Session Window, also reproduced on the next page. The columns represent gender and the rows represent the belief in the existence of Heaven. Each cell of the table contains a frequency count and a percentage. In the first cell for example, 107 of the 1224 respondents were women who did not say they believe in Heaven. They represented 8.74% of all respondents. The "All" row and column provide the *marginal* and relative frequencies for gender and belief.

```
Tabulated Statistics: HEAVEN2, SEX

 Rows: HEAVEN2      Columns: SEX

              Female      Male      All

 not Yes        107       129      236
               8.74     10.54    19.28

 Yes            568       420      988
              46.41     34.31    80.72

 Missing        116       105        *
                  *         *        *

 All            675       549     1224
              55.15     44.85   100.00
```

**11.**  *If we select a person at random, what is the probability that we select a woman? In other words, find P(F).*

**12.**  *If we select a person at random, what is the probability that we select a woman who believes in Heaven. (Note: this is the* **joint probability** *of F and H).*

**13.**  *Are men and women equally like to respond that they do believe that Heaven exists? Explain your thinking.*

This last question introduces us to the concept of *conditional* probability. In this sample, about 81% of all respondents expressed a belief in heaven. More specifically, 568 of the 657 females reported a belief in heaven. This works out to about 84% of the women. In comparison, only 420 of the 549 men (77%) responded similarly.

Imagine that we intend to randomly select one respondent and prepare to ask the question about belief in heaven. Prior to selecting one person, we would say that there is a probability of about 0.81 that the respondent believes in heaven.

Now let's imagine that we notice that the respondent is male. This fact should affect our estimate of the probability; statisticians say that the *conditional probability* of belief in heaven, *given a male respondent*, is 0.77.

The standard notation to express conditional probability is $P(BH|M)$, read as "the probability of event BH (believes in heaven) given that event M (male) has occurred." In this instance, we have two events in question and a marginal probability for each of them. Under the circumstances, or condition, that we know one of them has occurred, we often need to revise our assessment of probability for the other event.

To compute the conditional probabilities of belief in heaven for both genders, we can proceed as follows. Since gender is represented in the columns of this cross tabulation, we'll run the crosstab once more, this time computing column percentages rather than total percentages.

Edit the last dialog, unchecking **Total percents**, and instead checking **Column percents**.

14. **What is the probability that a respondent does not believe in heaven, given that the respondent is female? That is, find $P(D|F)$.**

## Independent Events

In this example, we just found that the men and women surveyed in the GSS differ in their beliefs. The women are more likely than the men to report a belief in heaven. If we ask ourselves how likely is it that a randomly chosen person expressed a belief in heaven, we might legitimately respond by saying that "it depends." The probability of the event in question seems to be linked to another event. In the language of probability, we would say that these events are not *independent.*

One of the most important concepts in the study of statistics is that of *independence.* If two events are independent, the occurrence of one does not affect the probability of the other. We generally says that if two events (A and B) are independent, then $P(A|B)=P(A)$. In English, this means that the conditional probability of A given B equals the marginal probability of A. If A and B are independent, then knowing that B has occurred does not effect the probability that A will occur.

## Moving On...

Learn what you have learned in this session to answer the following questions.

## GSSRelig

Recalling the events previously defined and also using the events M (person is male) and D (person does not believe in Heaven), find and interpret on the following probabilities:

1. P(H) = ?

2. P(D) = ?

3. P(M ∩ D) = P(M and D) =?

4. P(M ∪ D) = P(M or D) = ?

5. P(D|M) = ?

## New Worksheet

In a fresh worksheet, generate random data using the Integer Distribution (see page 76) with a minimum value of 0, and a maximum value of 1, as follows:

- In C1, generate 10 rows
- In C2, generate 100 rows
- In C3, generate 500 rows
- In C4, generate 1000 rows

6. Approximately what do you expect the mean value of each column to be, and why?

7. Compute the mean for each column. Comment on what you find in light of your answer to the previous question.

8. Now repeat the process of generating the four columns with four new sets of data, and compute the column means. Comment on how these results compare to your prior results. Why do the means compare in this way?

## Helpdesk

This worksheet contains data about user help requests received at a computer center help desk on a college campus.

9. If we were to randomly select a help request, what is the probability that the problem involved a hardware error?

10. If we were to randomly select one help request, what is the probability that the problem involved either a hardware error or a printer problem?

11. If we were to randomly select one help request, what is the probability that the problem involved both a network problem and an account problem?

## Tires

This worksheet contains data collected by the National Highway Safety Administration during July 2000. Each row refers to a complaint filed involving a tire failure problem.

12. Suppose you were to select one of these complaints at random; what is the probability that the car involved was manufactured by Ford?

13. What is the probability that the complaint involved a tread separation in the front driver-side tire?

14. What is the probability that the complaint involved a tread separation or that it involved the front driver-side tire?

15. If you knew that a complaint involved the front driver-side tire, what is the conditional probability that it also involved tread separation?

16. Based on your previous answers, would you conclude that tire position and the nature of the problem (e.g. tread separation) are independent? Explain.

## GSSSex2

This worksheet contains more data from the 1998 General Social Survey of adults in the United States. These questions all refer to sexual attitudes and behavior. *Note*: you may want to tally some of these variables before creating a cross tabulation, so that you can more easily read the categorical data values.

17. One might anticipate that the frequency of a respondent's sexual activity (**SEXFREQ**) is related to the respondent's marital status (**MARITAL**). Is it? Explain.

18. Is a respondent's attitude about homosexuality independent of the respondent's gender? Explain.

19. Respondents were asked if they had ever had sex with someone other than their spouse while married (**EVSTRAY**). Are responses independent of gender?

20. Given that a person is divorced, what is the probability that the person has ever strayed? What is the corresponding conditional probability for a person who is currently married? Comment on the comparison of these two conditional probabilities.

## Catalog2

This file holds a cross tabulation of data from an experiment conducted by a mail-order firm. The firm wants to know if small catalog changes can increase the probability that a customer will place an order.

In this case, three experimental catalogs were shipped along with a control ("Champion") catalog. In the control catalog, the company has a standard schedule of shipping and handling charges, as well as a small fee to insure the products in shipment. Additionally, customers normally have to judge the various available product colors by relying on the colored inks using in printing. The three experimental conditions were (1) reduced Shipping & Handling charges, (2) reduced shipping insurance charges, and (3) an offer of a free set of color samples with any order

21. Is the number of orders placed independent of catalog type?

22. What recommendations would you offer to the marketing director before the next catalog mailing?

## Violations

This worksheet contains records for 500 traffic violation citations issued at a small college.

23. Are residency status and class independent?

24. Are class and fine amount independent?

25. Are class and type of violation independent?

26. Given that a randomly selected citation was given to a Freshman, what is the probability that the fine was $10?

# Session 7

# Discrete Probability Distributions

## *Objectives*

In this session, you will learn to:

- Work with an observed discrete probability distribution
- Subset a worksheet to isolate selected observations
- Paste Session output into the worksheet
- Compute the expected value of a distribution
- Compute binomial probabilities
- Compute Poisson probabilities

## *An Empirical Discrete Distribution*

We already know how to summarize observed data; an *empirical distribution* is an observed relative frequency distribution which we intend to use to approximate the probabilities of a random variable. As an illustration, let's suppose that the chief of campus security in a small college is interested in the motor vehicle violations of resident students. We'll use the **Violations** data.

🖰 **File ➤ Open Worksheet...** Open the worksheet file **Violations**.

In this file, we are interested primarily in the variable called **Fine**, which represents the dollar value of a ticket issued to a motorist. Campus Security issues tickets in five different denominations. Therefore, **Fine** is a *discrete* variable. That is, it takes on a countable number of different numeric values.

The chief wants to focus on tickets issued to resident students on campus. Because the dataset also includes commuters, we need to extract a subsample from the entire dataset before analyzing the data.

🖰 **Data ➤ Subset Worksheet...** We'll create a new worksheet containing the same variables as **Violations**, but only for residents. The name of the new worksheet doesn't matter, but we must specify a **Condition** as part of the dialog.

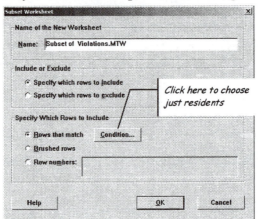

🖰 In the **Condition** dialog (below), type the expression **Res="R"**. This says that we want to include only those observations for residents. Click **OK**, and then **OK** in the main dialog.

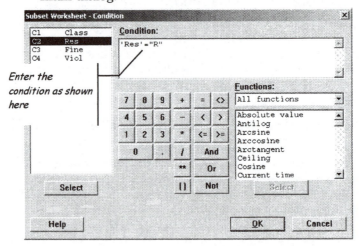

This creates a new worksheet, containing only the records pertaining to residents. Now we are ready to analyze the relative frequencies of the random variable, **Fine**.

🖰 **Stat ➤ Tables ➤ Tally Individual Variables…** Request **Counts** and **Percents** for the variable **Fine**.

1.   *In terms of probability, what do these percentages mean?*

| Fine | Count | Percent |
|------|-------|---------|
| 10 | 245 | 83.90 |
| 20 | 19 | 6.51 |
| 25 | 11 | 3.77 |
| 50 | 6 | 2.05 |
| 75 | 11 | 3.77 |
| N= | 292 | |

**Tally for Discrete Variables: Fine**

## Graphing a Discrete Random Variable

It is often helpful to graph a probability distribution, typically by drawing a line at each possible value of *X*. The height of the line is proportional to the probability. For this variable, let's begin with a dot plot.

🖰 **Graph ➤ Dotplot…** Select a simple dotplot, and choose the one available quantitative variable: **Fine**.

2.   *Comment on the shape of the distribution.*

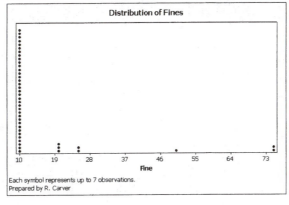

3.  *If we were to sample one ticket at random, what's the most likely fine amount?*

4.  *What was the average fine?  Which definition of "average" (i.e. measure of central location) is most appropriate here?*

## *Transferring Session Output to the Worksheet*

Suppose that we want to analyze the Tally results numerically. In particular, suppose we wish to compute the *expected value*[1] of the variable called **Fine**. We could type the tally results into the worksheet, but we can more easily copy the numbers from the Session Window, and paste them directly into the worksheet.

🖰  Scroll up in the Session Window so that you can see the entire tally. Position the cursor just to the left of the 10 in the `Fine` column, and click and drag the mouse, highlighting the entire numerical portion of the tally. Release the mouse button so that only the tally itself is highlighted.

**Tally for Discrete Variables: Fine**

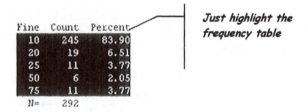

Just highlight the frequency table

```
Fine  Count  Percent
  10    245    83.90
  20     19     6.51
  25     11     3.77
  50      6     2.05
  75     11     3.77
 N=     292
```

🖰  After highlighting the relevant portion, select **Edit ➤ Copy**.

🖰  Move the cursor from the Session Window to Row 1 of an empty column in the Data Window.

🖰  **Edit ➤ Paste Cells...** In the message dialog that appears, indicate that spaces should be treated as delimiters.

---

[1] The *expected value* of a random variable, *X*, is the defined as $E(X) = \Sigma x_i f(x_i)$, where $f(x_i)$ is the probability that $X$ equals $x_i$. See the full discussion on the next page.

You should see the Frequency table (now a discrete distribution) appear in the worksheet. Label the three columns **X**, **f**, and **f(x)**.

## Computing the Expected Value of X

The expected value of a random variable, *X*, is its long-run mean. It is the sum of the products of each value of *X* times its probability. The formula is:

$$E(x) = \mu = \sum xf(x)$$

The next three commands find that value. We start by dividing f(x) by 100, to convert the values from percentages into probabilities.

> 🖱 **Calc ➤ Calculator...** As shown, divide f(x) by 100, and store the result back in the f(x) column.

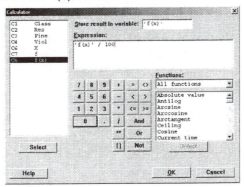

> 🖱 **Calc ➤ Calculator...** Using the first and third columns in the newly created portion of the worksheet, we'll create a new variable equal to *X* times *f(x)*. Store the result in a new variable called **Xf(x)**, and enter the **Expression** as **'X'*'f(x)'**.

> 🖱 **Calc ➤ Column Statistics...** Compute the **sum** of the new column. This is the *expected value* of X. Your input column is **Xf(x)**, and there is no storage variable.

> 🖱 **Calc ➤ Column Statistics...** Now compute the **mean** of **Fine**, and compare it to the expected value you just calculated.

Notice that, apart from a slight rounding difference, the sample mean and the expected value are the same. If you think about how we computed $f(x)$, this may make intuitive sense to you. In any event, it may serve as a good reminder that the expected value of a random variable is its long-run mean. In fact, we often denote $E(X)$ as μ.

## Cumulative Probability

Suppose the Chief of Campus Security was interested in knowing the percentage of all fines that exceeded $25. Referring to the earlier tally, we can easily see that 17 of the recorded fines, or just about 6%, were more than $25 in value.

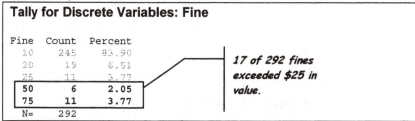

```
Tally for Discrete Variables: Fine

Fine   Count   Percent
  10     245    83.90
  20      19     6.51
  25      11     3.77
  50       6     2.05
  75      11     3.77
 N=      292
```

17 of 292 fines exceeded $25 in value.

Sometimes our questions about a set of data focus on a range of values, rather than on a single value. In such instances, it makes more sense to calculate *cumulative probability* rather than simple probability. We know that $f(x)$ represents the probability that the random variable, $X$, equals a particular value, $x$. In this book, we'll denote the *cumulative probability* as $F(x)$, which represents the probability that $X$ is less than or equal to a particular value $x$. Symbolically:

$$F(x) = P(X \le x)$$

With an empirical distribution, we can compute the cumulative probabilities by calculating the cumulative percentages (also known as the cumulative relative frequencies).

Stat ▶ Tables ▶ Tally Individual Variables... Earlier you requested counts and percentages. Now also request **Cumulative counts** and **Cumulative percents**.

5.   *Using the cumulative percents column in this tally, find an easy way to show that the probability that a randomly chosen fine exceeds $25 is approximately 0.06.*

6.   *What is the probability that a randomly chosen fine is less than $25?*

## *A Theoretical Distribution: The Binomial*

Some random variables arise out of processes which allow us to specify their distributions without empirical observation. *Binomial* random variables commonly arise in situations characterized by an iterative process that has two possible outcomes in each *trial*, or repetition, of the process. [2] Minitab can help us by simulating such variables or by computing their distributions. In this session, we'll focus on the latter function. We'll begin by computing the probability distribution for a binomial random variable with 8 trials and a probability of success ($\pi$) of 0.25 on each trial.

🖱 **File ➤ New...** Create a new Minitab Worksheet.

🖱 Into C1 of the blank worksheet, type the values 0 through 8. Label C1 **X** and C2 **f(x)**, as shown to the right:

| | C1 | C2 | C3 |
|---|---|---|---|
| ↓ | X | f(x) | |
| 1 | 0 | | |
| 2 | 1 | | |
| 3 | 2 | | |
| 4 | 3 | | |
| 5 | 4 | | |
| 6 | 5 | | |
| 7 | 6 | | |
| 8 | 7 | | |
| 9 | 8 | | |

🖱 **Calc ➤ Probability Distributions ➤ Binomial...** In this dialog, select **Probability**. Then specify that the **Number of trials** is **8**, the **Probability of success = 0.25**, the **Input column** is **X**, and the **Optional storage** column is **f(x)**. Click **OK**.

7. *What does this command do in the Data Window?*

🖱 **Graph ➤ Scatterplot...** Choose a simple scatterplot. Your **Y** variable is **f(x)** and the **X** variable is **X**. In the **Data View** subdialog, uncheck **Symbol** and check **Project lines**, and click **OK.**

8. *Comment on shape of distribution; where is its peak? Does the peak have any special significance?*

🖱 Title, label, and print this graph.

🖱 **Calc ➤ Probability Distributions ➤ Binomial...** Change the **Probability of success** to **0.4** and **Optional storage** to **C3**.

---

[2] In this section, we assume that the reader has some familiarity with binomial distributions. Consult your primary textbook for a description of the conditions under which a binomial random variable arises.

9. *Comment on differences between C2 and C3. How will the graph of C3 compare to that of C2?*

We can also compute and work with cumulative binomial probabilities. In our first binomial example, we imagined an experiment with eight trials and a 0.25 probability of success. Suppose now that this example referred to the following situation. A market researcher has divided a company's client base into several geographical regions, and 25% of the clients are in the Western region.

Now the researcher plans to contact, at random, eight independently-selected clients. Thus we have eight identical, independent trials, and $\pi = 0.25$ for each trial. Suppose further that the researcher wishes to know the probability of obtaining *fewer than* two Western clients. In other words, she wants to find $F(8)$ rather than $f(8)$.

Calc ➤ Probability Distributions ➤ Binomial...   First, select **Cumulative** in the upper right of the dialog. Change the **Probability of success** back to **0.25** and **Optional storage** to **C4**.

The cumulative probabilities are now in Column 4, which you might want to label **F(X)**.

10. *What is the probability that fewer than two of the clients will be from the West?*

11. *What is the probability that at least one client will be from the West?*

We can use this line of reasoning to develop a binomial model that describes a real-world situation. In the General Social Survey data, it turns out that just about 5% of the respondents were from New England. Suppose we want to select a 20-person random subset of these respondents, and want to know the probability that at least two people in such a subset came from New England.

Before doing any computation, let's think this through. We have a binomial random variable with $n = 20$ and $\pi = 0.05$. We want to find the probability that the variable exceeds 1; thus we want to find $P(X > 1)$.

12. *Explain or show why P(X>1) is the complement of F(1).*

Much as you did above, generate the cumulative binomial probabilities using the parameters $n = 20$ and $\pi = 0.05$.

13. *What is the probability that a randomly-selected sample of 20 respondents will include at least two from New England?*

**14.**   *What is the probability that a randomly-selected sample of 20 respondents will include at most two from New England?*

## Another Theoretical Distribution:  The Poisson

We can compute several common discrete distributions besides the Binomial. Let's look at one more. The *Poisson* distribution is often a useful model of events that occur at random over a fixed period of time or distance. [3] The distribution has but one parameter, and that is its mean. Using the same worksheet as for the binomial example, do the following:

🖱  **Calc ➤ Probability Distributions ➤ Poisson...** In this dialog, specify a **Mean** value of **2**, and **Input column** of **X**, and an **Optional storage** column of **f(x)**.

🖱  Plot this variable as you did with the binomial, and print the graph.

**15.**   *How do these graphs compare?*

We can use the theoretical Poisson distribution as a model of a real-world process, assuming that the process is one in which we can observe the occurrence of some event over time or distance. For example, consider the number of automobile accidents experienced by a college-aged driver during a given period of time. The number of accidents is countable, occurring over a continuous time period. We might wonder whether a Poisson distribution could accurately describe the relative frequency of accidents.

Open the **Student** worksheet file.  One of the survey questions asked these students how many automobile accidents they had been involved in during the previous two years.

🖱  **Stat ➤ Tables ➤ Tally Individual Variables...** Tally the variable **Accident**. Note that within this particular sample, no student had more than 3 accidents; naturally, it is conceivable that a respondent might have had more than 3 such incidents.

Let's construct a theoretical Poisson distribution with a mean equal to the sample mean of **Accident**. In this way, we can see if the

_____

[3] As with the discussion of binomial probabilities, we assume familiarity with the Poisson distribution.

theoretical distribution might be a useful generalization from this single sample of students.

We can easily compute the sample mean; confirm for yourself that it equals 0.654545 accidents per student in the prior two years.

🖱 Now let's repeat the Poisson computation from the previous example. Use the **Window** menu to re-activate the worksheet containing **X** and **f(x)**.

🖱 **Calc ➤ Probability Distributions ➤ Poisson…** Just change the mean to **0.654545**, and click **OK**.

The **f(x)** column now contains the probabilities for a theoretical Poisson distribution with a mean of 0.654545. Notice how well these proportions compare to the tabulated percentages from our student dataset. Though not exactly the same, each theoretical percentage is within a few points of the observed relative frequencies. Thus, if we simply know that the mean number of accidents if approximately 0.65, we can conveniently use the Poisson distribution as a good first approximation to describe the shape of the random variable.

## Moving On…

Let's use what we have learned to (a) analyze an observed distribution and (b) see how well the Binomial or Poisson distribution serves as a model for the observed relative frequencies.

### Student

Students were also asked how many siblings they have. If we were to assume that most women bear children during a similar period of their lives, we might wonder if the Poisson distribution can accurately model the number of siblings a student reports. The variable called **"Sibling"** records their answers. Perform these steps to answer the question below:

a)  Tally the number of siblings, computing the relative frequencies.
b)  Find the mean of this variable (equals the expected value)
c)  In an empty column of the worksheet, type the values 0 through 9 (i.e., 0 in Row 1, 1 in Row 2, etc.)

d) Generate a Poisson distribution with a mean equal to the mean number of siblings. The input column is the one you just typed, and the output column is the next one.

e) Type the computed relative frequencies (as proportions, not percentages) into the worksheet in the column adjacent to the Poisson probabilities.

1. Compare your tally of actual accidents to the Poisson distribution (either visually or graphically). Does the Poisson distribution appear to be a good model or approximation of the actual data?

2. Considering your answer to the prior question, suggest a reason that might explain your observed "fit" between the Poisson model and the actual data.

## Pennies

A professor has his students each flip 10 pennies, and record the number of heads. Each student repeats the experiment 30 times and then records the results in a worksheet.

3. Compare the actual observed results (in a graph or table) with the theoretical Binomial distribution with $n = 10$ trials and $p = 0.5$. Is the Binomial distribution a good model of what actually occurred when the students flipped the pennies? Explain. (Hint: First, find the mean of each column; since each student conducted 30 experiments, the mean should be approximately 30 times the theoretical probability.)

4. According to the theoretical distribution, what is the probability of flipping at least 4 heads in 10 trials? According to the actual student data, in what percentage of all experiments were there at least 4 heads in 10 trials?

## Web

Twenty trials of twenty random queries were made using the Yahoo!® Internet search engine's Random Yahoo! Link. For some links, instead of successfully connecting to a Web site, an error message appeared. In this data file, the variable called **problems** indicates the number of error messages received in each set of twenty queries. Perform the following steps to answer the questions below:

a) Find the mean of the variable **problems** and divide it by 20. This will give you a percentage, or probability of success (obtaining an error message in this case) in each query.

b) Create a new variable **prob** (for number of possible problems encountered) and type the values **0** through **20** (i.e. 0 in Row 1, 1 in Row 2, etc.)

c) Label another new variable column as **binom**.

d) Generate a theoretical cumulative Binomial distribution with $n$= 20 (number of trials) and a probability of success equal to the proportion you found in step (a).

e) Using the Tally command, produce a cumulative relative frequency distribution for the variable **problems**.

5.  Compare the actual cumulative percent of problems to the theoretical Binomial distribution. Does the Binomial distribution provide a good approximation of the real data? Comment on the similarities and differences as well as reasons they might have occurred.

6.  Using this theoretical Binomial probability, what is the probability that you will receive exactly three error messages? How many times did this actually occur? Why are there differences? If the sample size were $n = 200$, how large do you think the differences would be?

## GSSGeneral

This file contains responses from 1445 individuals participating in the General Social Survey. Respondents are identified by geographical region. For these questions, we'll treat these 1445 people as a population.

7.  Which region has the highest percentage of respondents? Explain your choice.

8.  If we were to randomly select a sample of 30 respondents from this population of 1445 individuals, what is the expected value of the number of people from that region within the sample of 30?

9.  What is the probability that a random sample of 30 people would have no one from that region?

What is the probability that a random sample of 30 would have 5 or more people from that region?

# Normal Distributions

## *Objectives*

In this session, you will learn to:
- Create a column of "patterned data"
- Compute probabilities for any normal random variable
- Use normal curves to approximate other distributions

## *Continuous Random Variables*

The prior session dealt exclusively with discrete random variables, that is, variables whose possible values can be listed (like 0, 1, 2, etc.). In contrast, some random variables are *continuous*. The defining characteristic of a continuous random variable is that, for any two values of the variable, there are an infinite number of other possible values between them. Such variables cannot be tabulated as can discrete variables, nor can a unique probability be assigned to each possible value. As such, we must think about probability in a different way when we work with continuous random variables.

Rather than constructing a probability distribution, as we did for discrete variables, we think in terms of a *probability density function* when dealing with a continuous random variable, $X$. We'll envision probability as being diffuse over the permissible range of $X$; sometimes the probability is "dense" near particular values, meaning that these values occur frequently. The density function itself is difficult to interpret, but the area beneath the density function (i.e. its *integral*) represents probability.

Perhaps the most closely studied family of continuous random variables is the *normal distribution*. We begin this session by considering several specific normal random variables.

## *Generating Normal Distributions*

There are an infinite number of normally distributed random variables, each with its own mean and standard deviation. The mean and standard deviation, then, are the *parameters* of the distribution. If we know that $X$ is normal, with mean $\mu$ and standard deviation, $\sigma$, we know all there is to know about $X$. Throughout this session, we'll denote a normal random variable as $X \sim N(\mu, \sigma)$. We read this notation as "X is distributed normally with a mean of $\mu$ and a standard deviation of $\sigma$. For example, $X \sim N(10,2)$ refers to a random variable $X$, which is normally distributed with a mean value of 10, and a standard deviation of 2.

Although each member of the normal family has a unique mean and standard deviation, all normal distributions have certain features in common. For example, all normal density curves are perfectly symmetric, forming a bell shape. All normal curves share the same proportional relationships between distances on the horizontal axis and areas under the curve. For example, nearly 100% of the area is accounted for in the range of six standard deviations: from three standard deviations below the mean to three standard deviations above the mean.

Before learning how to find probabilities associated with normal variables, we'll consider the density function for three different distributions, to see how the mean and standard deviation define a unique curve. Specifically, we'll generate values of the density function for a *standard normal variable*, $z \sim N(0,1)$, and two others: $X \sim N(1,1)$ and $X \sim N(0,3)$.

   ⌐🖰 In the Data Window, label the first variable (column 1) **X**, label **C2** as **Z** (representing the standard normal variable), **C3** as **N11** and **C4** as **N03**.

   ⌐🖰 **Calc ➤ Make Patterned Data ➤ Simple Set of Numbers...** In **C1**, we will generate a column of values ranging from -8 to +8 with an increment of 0.2. We could type all of the values into the column, but this command does it for us.

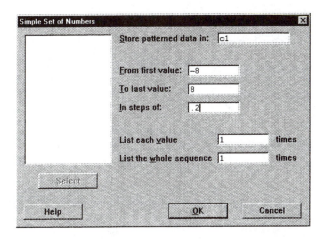

🖱 **Calc ➤ Probability Distributions ➤ Normal...** This command will compute values of the normal density function. Click the button marked **Probability density**. Specify a **Mean = 0** and **Standard deviation = 1**. The input column is **X**, and the **Optional Storage** column is **Z.** Now C2 contains density values for $X \sim N(0,1)$, which is the standard normal curve.

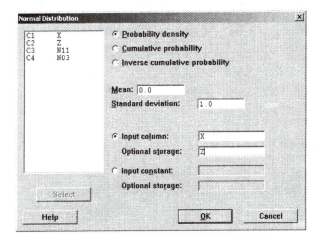

🖱 **Edit ➤ Edit Last Dialog...** Change the **Mean** to **1**, and the **Optional storage** column to **N11**.

🖰 Edit the last dialog once more, this time changing the mean to 0, the standard deviation to 3, and the optional storage column to **N03**.

1. *If you were to graph these three normal variables, how would the graphs compare?*

2. *Where would the curves be located on the number line?*

3. *Which would be steepest and which would be flattest?*

Let's create a graph of the three curves and see how they compare.

🖰 **Graph ➤ Scatterplot...** In the first dialog, select **With Connect and Groups**.

🖰 In the main dialog box, specify the variables for three graphs. The Y variables are **Z**, **N11**, and **N03**. For each graph, the X variable is just **X**. Be sure the check the box marked **X-Y pairs form groups**, as shown here.

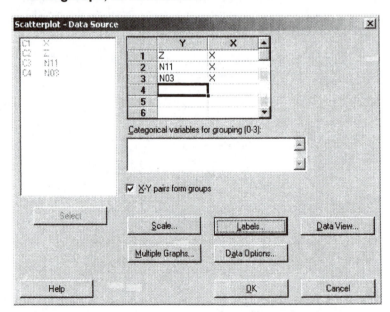

🖰 Title and place your name on the graph (click **Labels**).

Look closely at the resulting graph (see next page).

4.  *Describe what you see in the graph: How do these three normal distributions compare to one another?*

5.  *How do the two distributions with a mean of 0 differ?*

6.  *How do the two with a standard deviation of 1 differ?*

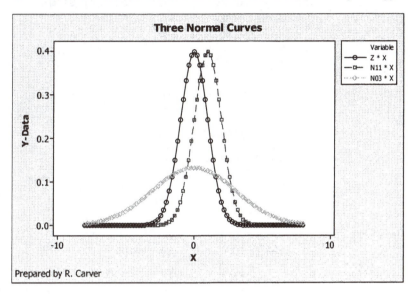

Print this graph.

## Finding Areas Under a Normal Curve

There are few practical applications of the normal density function itself, but we often need to compute the probability that a normal variable lies within a given range. Traditionally—without software—one would convert a variable to the standard normal variable, $z$, consult a table of values, and then manipulate the areas to find the probability. With Minitab, we can find such probabilities easily using the cumulative probability function. Recall that the *cumulative probability* of $x_i$ is the probability that $X$ is less than or equal to $x_i$. Therefore, if we want to find the probability that $X$ lies between two values (say $a$ and $b$), we'll just find the cumulative probability corresponding to each value and subtract. Here is an example using the standard normal variable, $z$.

Suppose we want to find the probability that an observation lies within one standard deviation of the mean in a standard normal distribution. That is, we want to find P($-1 \leq z \leq +1$).

🖰 **Calc ➤ Probability Distributions ➤ Normal...** At the top of the dialog box, click **Cumulative Probability**. The values generated represent the area under the curve from $-\infty$ to the specified value in the **Input Column**. Set the **Mean = 0** and **Standard Deviation = 1**. Click the radio button marked **Input Constant**. Type **–1** in the **Input Constant** box, and click **OK**.

🖰 Repeat this process, this time entering **1** in the **Input Constant** box.

In the Session window, you should see the following:

```
Cumulative Distribution Function
Normal with mean = 0 and standard deviation = 1

  x   P( X <= x )
 -1      0.158655

Cumulative Distribution Function
Normal with mean = 0 and standard deviation = 1

  x   P( X <= x )
  1      0.841345
```

Thus, we see that P ($z \leq -1$) =.1587 and P ($z \leq 1$) = .8413. As noted earlier, the probability we want is just the difference between the two, or .8413 – .1587 =.6826. You may have learned that 68% of the area for a normal curve lies within one standard deviation of the mean. Now you know where the 68% figure originates.

We can also find several cumulative probabilities at once by specifying an input column rather than an input constant. Suppose we want to find P ($-2.5 \leq z \leq 2$). We could edit the Normal Probability dialog twice, entering each value separately, or we could request them directly, as follows.

🖰 In empty column (C5), type the two values **2** and **–2.5**.

🖰 **Calc ➤ Probability Distributions ➤ Normal...**  In the lower portion of the dialog, click the radio button marked **Input Column** and specify **C5** as the input column. Then choose **C6** as **Optional storage** column.  After clicking **OK**, you'll see this in the worksheet:

| ↓ | C1 | C2 | C3 | C4 | C5 | C6 | C7 |
|---|----|----|-----|-----|----|----|----|
|   | X  | Z  | N11 | N03 |    |    |    |
| 1 | -8.0 | 0.000000 | 0.000000 | 0.003799 | 2.0 | 0.977250 |   |
| 2 | -7.8 | 0.000000 | 0.000000 | 0.004528 | -2.5 | 0.006210 |   |
| 3 | -7.6 | 0.000000 | 0.000000 | 0.005373 |   |   |   |

Subtract the two values in C6 for the desired result. The diagram below illustrates what we have done. Recall that probability is determined by area under the curve, and that the entire area under the curve equals 1.0. In the curve shown below, the shaded area represents $P(-2.5 \leq z \leq 2)$; when we subtract the two cumulative probabilities, we are finding the shaded area.

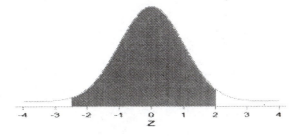

This approach works for any normally distributed random variable. Suppose X is Normal with a mean of 500 and a standard deviation of 100. Let's find $P(500 < X < 600)$.

🖱 Type **500** and **600** into top two cells of **C5**.

🖱 Edit the last command dialog again. Change the **Mean** to **500**, the **Standard Deviation** to **100**. We can retain columns 5 and 6 for input and output. Once again, subtract the two result values.

7. *Now find $P(X \leq 300)$ for $X \sim N(500,100)$.*

8. *Find $P(X \geq 600)$ for the same distribution (think about how you want to proceed here).*

## Inverse Normal Probabilities

Now we know how to compute the probability that a normal variable will take on values in a specific range on the number line. In other words, given a location on the horizontal axis, we can find the corresponding probability. In some situations, we actually need to do the inverse operation: We need to find a location on the axis corresponding to a predetermined probability.

For example, we learned earlier that nearly the entire curve lies within three standard deviations of the mean. Let's locate the middle 50% of a normal curve. In other words, let's ask "what are the axis values that represent the first and third quartiles of a normal curve?" Let's return to the standard normal curve for this first illustration. Algebraically we want to find the particular values $z_1$ and $-z_1$ such that $P(-z_1 \leq z \leq z_1) = 0.5000$. That is to say, we want $P(z \leq -z_1) = .25$ and we also want $P(z \leq z_1) = .75$. Graphically, our task is this:

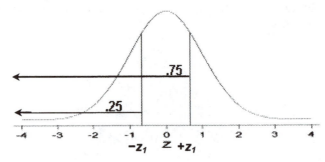

🖰 In the worksheet window, replace the earlier contents of **C5** with the values **.75** and **.25**.

🖰 Return to the Normal Probability dialog, and select **Inverse cumulative probability**. Specify a mean of **0** and a standard deviation of **1**; be sure that you have selected **Input column** equal to **C5** and **Optional Storage** in **C6**.

Find the results in the Session window. You should find that the quartiles occur at $z = \pm 0.674490$. In round numbers the first and third quartiles fall at approximately 2/3 of one standard deviation above and below the mean. In other words, the middle 50% of the curve lies within 2/3 of a standard deviation of the mean.

9. *Consider a variable, X~N(100,30). Based on the knowledge that the middle 50% of the curve lies within 2/3 of a standard deviation, estimate the first quartile for X.*

10. *Now, use Minitab to find the first quartile for this variable.*

**11.   Find −X₁ and +X₁ such that P(−X₁ ≤ X ≤ +X₁) = .95. Comment on your finding. (Hint: make a sketch to represent your goal in this question.)**

## Normal Curves as Models

One reason that the normal distribution is so important is that it can serve as a close approximation to a variety of other distributions. For example, binomial experiments with many trials are approximately normal. Let's try an example of a binomial variable with 100 trials, and $\pi$ = P(success) = 0.20.

🖱 **Calc ➤ Make Patterned Data ➤ Simple Set of Numbers...** In C8, generate rows from 0 to 100 in steps of 1. Click **OK**.

🖱 **Calc ➤ Probability Distributions ➤ Binomial...** Choose **Probability** and set the number of trials = 100, P(success) = 0.2 The **Input column** is **C8**, and **Optional storage** is **C9**. C9 now contains the Binomial probability values for the number of successes in 100 trials.

🖱 **Graph ➤ Scatterplot...** Create a simple scatterplot of C9 vs. C8

**12.   Do you see that this distribution could be approximated by a normal distribution?**

The question is: "*which* normal distribution in particular?" Since $n$ = 100 and $\pi$ = .20, the mean and standard deviation of the binomial variable are 20 and 4.[1] Let's generate a normal curve with those parameters.

🖱 **Calc ➤ Probability Distributions ➤ Normal...** Click **Probability Density**. The **Mean** is **20** and the **Standard Deviation** is **4**. The **Input column** is **C8**, **Optional storage** is **C10** (which is currently empty).

🖱 Create another simple scatterplot of c10 vs. c8 on the same axes as c9 vs. c8 (**Multiple Graphs; Overlaid on the same page**).

---

[1] For a binomial X, $E(X) = \mu = np$. Here, that's (100)(.20)=20. Similarly, the standard deviation is $\sigma = \sqrt{np(1-p)} = \sqrt{100 \cdot .20 \cdot 80} = \sqrt{16} = 4$.

**13.** *Comment on the comparison between the two curves; what advantages and disadvantages might we realize if we use the normal curve to approximate the binomial probabilities?*

The normal curve is also often a good approximation of real-world observed data. Let's consider an example.

🖰 **Open** the file **Sleep**, which contains data about the sleeping habits of 62 mammalian species. We'll focus on the average number of hours of non-dreaming sleep for each species.[2]

🖰 **Graph ➤ Histogram…** In the opening dialog, select **With Fit**. The select the variable **Sleepnon**. This will create a histogram with a superimposed normal curve. The parameters of normal curves are the sample mean and standard deviation for our variable. In other words, Minitab constructs a theoretical normal curve using our sample mean and standard deviation.

**14.** *In your judgement, how closely does the normal curve approximate the histogram?*

We can go a step beyond a simple visual inspection to make some judgments about how closely the normal curve matches the empirical distribution of our sample. We know that the normal curve and our sample data share the same mean and standard deviation, because Minitab did that deliberately. Now let's see how the empirical distribution and the theoretical normal distribution match up at five other values: the minimum, first quartile, median, third quartile, and maximum. Recall that these values are known as the *five number summary*.

🖰 Obtain the five number summary either by generating the standard descriptive statistics or a graphical summary for **Sleepnon**.

Let's now turn our attention to the theoretical normal distribution whose mean and standard deviation match those of **Sleepnon**. From earlier in our session, we know how to find the quartiles. We'll do that again now.

🖰 In the data window, move to an empty column (like C12) and type in the values .25, .50, and .75 corresponding to each of the quartiles and the median.[3]

---

[2] Sleeping time can be divided into dreaming and non-dreaming sleep.

🖰 **Calc ➤ Probability Distributions ➤ Normal...** We want to compute inverse cumulative probabilities for the normal distribution whose mean is 8.67292 and whose standard deviation is 3.66645. The input column is C12 (where we entered the percentile values above), and we want to store the results in the next column (C13).

This yields three values: 6.1999, 8.6729, and 11.1459. These are the three quartile figures. As a first step in comparing the observed data values with a corresponding theoretical normal distribution, we can compare the quartiles are benchmarks. If the observed quartile values are approximately equal to the theoretical quartiles, it may be appropriate to use the normal curve as an approximation for the actual non-dreaming sleep times. The table below summarizes our calculations, comparing the observed (empirical) figures with our theoretical figures.

|  | Observed Sleepnon | Theoretical |
|---|---|---|
| 1st Quartile | 3.15 | 6.20 |
| Median | 8.35 | 8.67 |
| 3rd Quartile | 11.00 | 11.15 |

We can scan down the two columns and note that the 1st quartile of the empirical distribution is rather different from the theoretical distribution, but the others are really quite close. In other words, we can begin to think about the degree to which this set of data is well approximated by a normal curve. Just as we can closely examine three particular values from the dataset, we can compare all of the observations to a theoretical normal curve using a *normal probability plot*.

## Normal Probability Plot

A probability plot is a scatter plot that compares observed and theoretical values much in the same manner that we have just described.[4] On the vertical axis we find percentile values for a theoretical normal distribution sharing the same mean and standard deviation as the empirical data. The actual data values appear on the horizontal axis. If the two distributions matched perfectly, plotted points would fall on a

---

[3] Without any computation at all, we know that the median will equal the mean value of 8.67292 hours of non-dreaming sleep, but it won't hurt to compute it.

[4] Different authors and software packages adopt slightly different conventions in constructing normal probability plots. This describes Minitab's approach. The description here is simplified for the introductory reader. See Minitab's own documentation for a fuller explanation.

straight line with a slope of 1.0—a line rising from the origin at a 45°
angle. Let's construct the plot for **Sleepnon**.

⌐🖰   **Graph ➤ Probability Plot...** Choose a **Single** plot, and select the
variable **Sleepnon** and click **OK**.

The command generates some summary statistics, a table and a
graph. At this stage of our discussion, we'll only focus on the graph. In
Session 10, we'll return to the summary statistics. Here is the graph for
our variable:

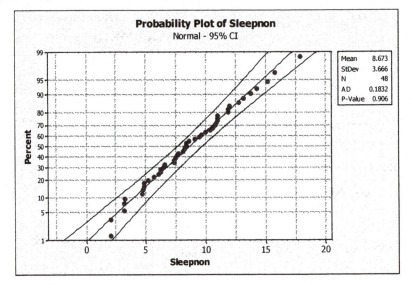

Here we see that, for the most part, the points wobble around the
diagonal line with a few exceptions. As a generalization, though, it would
be fair to say that these points form an approximate 45° straight line.
Accordingly, we can be relatively confident that the normal curve is a
good approximation of this dataset. At this point in your course, you can
follow this general guideline: if all of the plotted points fall within the two
curved bands above and below the 45° line, you are looking at an
approximately normal distribution.

In contrast, construct a normal probability plot for the variable
called **Weight**, which represents the mean body weight (in kilograms) for
the species. In addition, create a Graphical Summary using the
Descriptive Statistics command.

**15.   *Describe the normal probability plot that you
constructed; do the points fall in a 45° straight line?***

16. *In the graphical summary, do the sample data appear to follow a normal distribution?*

## Moving On...

Use what you have learned in this session to complete the following exercises.

**New Worksheet** (or empty columns of another worksheet)

1. Consider a theoretical normal random variable, $X$, with a mean of 8 and a standard deviation of 2.5. Compute the following probabilities:

   - $P(7 \leq X \leq 8.5)$
   - $P(9 \leq X \leq 10)$
   - $P(X \geq 4)$
   - $P(X < 4)$
   - $P(X > 10)$

2. Into a blank worksheet column, generate a simple set of numbers from 0 to 200. In the adjacent column, generate binomial probabilities for a binomial distribution with $n = 200$ and $\pi = 0.4$. In the next column, compute the appropriate normal probability densities. Construct a graph to compare the binomial and normal probabilities; comment on the comparison.

3. Consider a theoretical normal random variable, $X \sim N(25,3)$. Find inverse cumulative values of $x^*$ such that:

   - $P(X < x^*) = .05$
   - $P(X > x^*) = .10$
   - $P(x_1^* < X < x_2^*) = .90$ ($x_1^*$ and $x_2^*$ equidistant from 25.)
   - $P(x_1^* < X < x_2^*) = .50$ ($x_1^*$ and $x_2^*$ equidistant from 25.)

**Output**

This file contains monthly data about the industrial output of the United States for many years. The first column contains the date, and the next six contain specific variables described in Appendix A.

Using the **Display Descriptive Statistics** command, generate a histogram with normal curve superimposed for *all six variables*.

4.  Based on their histograms, which of the six variables looks most nearly normally distributed to you? Least nearly normal?

5.  For the variable you chose as most nearly normal, compute the quartiles (including the median) for the appropriate theoretical normal distribution. Discuss the comparison of your theoretical values and the corresponding values from the dataset.

6.  Construct normal probability plots to confirm your judgment for these two variables, and explain what you find.

7.  Suggest some "real world" reasons that the variable you selected as most nearly normal might follow a normal distribution.

## Bodyfat

This file contains body measurements of 252 men. Using the **Display Descriptive Statistics** command, generate a histogram with normal curve superimposed for *all* of these variables:

- FatPerc
- Age
- Weight
- Neck
- Biceps

8.  Based on their histograms, which of the variables looks most nearly normally distributed to you? Least nearly normal?

9.  Suggest some "real world" reasons that the variable you selected as most nearly normal would follow a normal distribution.

10. For the neck measurement variable, what are the parameters (mean and standard deviation) of a normal curve, which closely approximates the observed data?

11. Using the approximate normal distribution that you just identified, compute the theoretical first and third quartiles. Compare these theoretical quartiles with the

actual quartile values for the observed neck measurements.

12. Use Minitab's cumulative normal command and the normal distribution just identified to estimate the percentage of men with neck measurements between 29 and 35 cm.

## Water

These data concern water usage in 221 regional water districts in the United States for 1985 and 1990. Compare the normal distribution as a model for **C17** and **C34** (you investigated these variables earlier in Session 4).

13. Which one is more closely modeled as a normal variable?

14. What are the parameters of the normal distribution which closely fits the variable **Percentcu** (C34)?

15. Construct a normal probability plot for **Percentcu** and comment on what you see.

16. What concerns might you have in modeling **Percentcu** with a normal curve? (Think about the theoretical maximum and minimum of a normal curve.)

## MFT

This worksheet holds scores of 137 students on a Major Field Test, as well as their GPA's and SAT verbal and math scores.

17. Construct a normal probability plot for the math SAT scores; do they appear to be normally distributed?

18. Identify the parameters of a normal distribution, which closely approximates the Math scores of these students.

19. Use Minitab's cumulative normal command and the distribution you have identified to estimate how many of the 137 students scored more than 59 on the Math SAT.

20. We know from the Descriptive Statistics command that the Third Quartile (75 percentile) for Math was 59. How can we reconcile your previous answer and this information?

## MCASELA

This worksheet holds the results of the English Language Arts portion of Massachusetts Comprehensive Assessment System (Grade 4) exams for 2000. Each row represents one elementary school. Students are scored as *Advanced, Proficient, Needs Improvement,* or *Failing,* and the percentage of students at each of these four levels is reported for each school.

21. Construct a histogram of the variable **niper** (percentage of students at the Needs Improvement level). Comment on the extent to which it appears to be normally distributed.

22. Compute the mean, standard deviation, median, and quartiles for **niper**. Then compute the median and quartiles for a theoretical distribution whose mean and standard deviation are the same as **niper**. Compare the medians and quartiles of the empirical and theoretical distributions; which of the pairs of values of most similar? Which are most dissimilar? Looking at the histogram, does this make sense to you? Discuss.

23. Construct a histogram of the variable **mnscore** (mean score of all students in the school). Comment on the extent to which it appears to be normally distributed.

24. Now construct a normal probability plot for this variable. Comment on the extent to which it appears to be normally distributed.

# Session 9

## Sampling Distributions

### *Objectives*

In this session, you will learn to:
- Simulate random sampling from a known population
- Use simulation to illustrate the Central Limit Theorem

### *What is a Sampling Distribution?*

Every random variable has a probability distribution or a probability density function. One special class of random variables is that *of statistics computed from random samples.*

How can a statistic be a random variable? Consider a statistic like the sample mean $\bar{x}$. In a particular sample, $\bar{x}$ depends on the $n$ values in the sample; a different sample would potentially have different values, probably resulting in a different mean. Thus, $\bar{x}$ is a quantity that varies from sample to sample, due to the luck of random sampling. In other words, it's a quantitative random variable.

Every random variable has a distribution with shape, center and dispersion. The term *sampling distribution* refers to the distribution of a sample statistic. In this session, we'll simulate drawing many random samples from populations whose distributions are known, and see how the sample statistics vary from sample to sample. Naturally, this is unrealistic. Our goal is to gain insight into the extent of variation introduced through the process of random sampling.

### *Sampling from a Normal Population*

We start by considering a large sample from a population known to be normally distributed, with $\mu = 500$ and $\sigma = 100$.

🖰 **Calc ➤ Random Data ➤ Normal**  Generate 60 rows in Column 1, mean = 500, standard deviation = 100, as shown here:

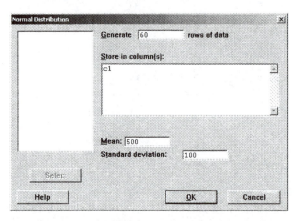

  Look at C1 in your Data Window. Remember that it's a random sample, different from your neighbors' and different still from another sample you might have drawn. The question is: "How much different?"

  From what you know about normal populations, you can anticipate that the mean of this sample will be close to 500, and that about 95% of the sample values should lie within two standard deviations of the mean. In other words, nearly all of you sample values should be in the range of 300 to 700. In this session, though, we want to focus our attention on the sample mean, and specifically how the value of the sample mean might fluctuate from sample to sample.

  Since the mean of the population is 500, it is reasonable to expect the mean of this first column to be near 500. It may or may not be "very" near, but the result of one simulation doesn't tell us much. To get a feel for the randomness of $\bar{x}$ , we need to consider many samples.

  Let's think of C1 as containing the first observation ($x_1$) of 60 simulated random samples. Let's see what happens if we simulate 60 samples, each with $n$ = 80 observations. We'll proceed to generate 60 rows in the next 79 columns of the worksheet. Then we can compute the sample means for each of the 60 samples, create a column containing all 60 sample means, and then make some comparisons among them.[1]

---

[1] It may seem more "natural" to treat the columns as samples, since we usually think of observations as occupying rows. Because of the simplicity of computing *and storing* the means of rows (but not columns) in Minitab, we'll proceed as described in this session.

🖰 **Edit ➤ Edit last dialog...** In the **Store in column(s)** box, change **C1** to **C2-C80.** Click **OK**.

🖰 Move to the Data Window, and label C81 as **Mean**.

🖰 **Calc ➤ Row Statistics...** Select **Mean** of **C1–C80**, and store the result in **C81**, as shown here:

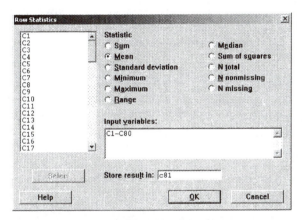

Recall that each *row* of this worksheet is a random sample of $n = 80$ observations, so C81 now contains the means of all 60 samples; we can consider it a random variable, since each sample mean is different due to the chance involved in sampling.

**1.   *What should the mean of these sample means be?***

🖰 **Stat ➤ Basic Statistics ➤ Display Descriptive Statistics...** Select **C81**.

Despite the fact that your random samples are unique and individually unpredictable, we can confidently predict that the mean of **C81** will be very nearly 500. This is a key reason that we study sampling distributions. We can make very specific predictions about the sample mean in repeated sampling, even though we cannot do so for one sample.

How much do the sample means vary around 500? In a random sample from an infinite population, *the standard error of the mean* is given by this formula:

$$\sigma_{\bar{x}} = \frac{\sigma}{\sqrt{n}}$$

In this case, $\sigma = 100$ and $n = 80$. So here,

$$\sigma_{\bar{x}} = \frac{100}{\sqrt{80}} = \frac{100}{8.944} = 12.3$$

2.   *Find the standard deviation of C81 in the Session Window. How close is it to 12.3?*

Remember that the standard error is the theoretical standard deviation of all possible values of $\bar{x}$, and the standard deviation of C81 represents only 60 of those samples.

So far we've looked at the center and spread of the distribution of $\bar{x}$; what about its shape?

🖱 **Edit ➤ Edit last dialog...** Choose **C80** and **C81**. Before clicking **OK**, click on **Graphs**, and choose **Histogram of data, with normal curve**.

🖱 **Editor ➤ Layout...** The Layout screen allows you to arrange multiple graphs for comparison. Your screen should look like this, though your histograms will be unique.

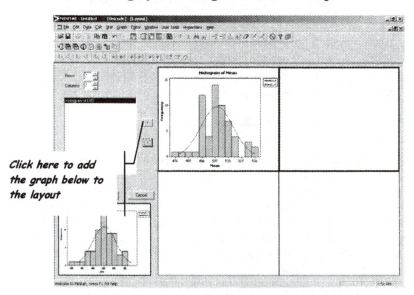

*Click here to add the graph below to the layout*

The default layout has two rows and two columns, which you could adjust using the controls in the upper left of the screen. We can leave the default setting for now. Notice that the quadrant in the upper right of your screen is outlined in red. This is the target for another graph. Click on the right-facing arrow, as shown, to move the second graph into the layout, so that you can compare these two distributions.

3. *How would you describe the shapes of the graphs?*

4. *What do you notice about their respective centers and spread? (Look closely at the horizontal axis).*

5. *Suggest a possible reason for the comparison of the range of these two distributions: why are they so different?*

## Central Limit Theorem

The histogram of C81 was roughly normal, describing the means of many samples from a normal population. That may seem reasonable: The means of samples from a normal population are themselves normal. What about samples from non-normal populations?

According to the Central Limit Theorem, the distributions of sample means approach a normal curve as $n$ grows large, *regardless of the shape of the parent population*. To illustrate, let's take 60 samples from a uniform population ranging from 0 to 100. In a *Uniform* population with a minimum value of $a$ and a maximum value of $b$, the mean is found by:

$$E(X) = \mu = \frac{(a+b)}{2}$$

In this population, that works out to a mean value of 50. Furthermore, we can compute the population variance as:

$$Var(X) = \sigma^2 = \frac{(b-a)^2}{12}$$

In this population, the variance is 833.33, and therefore the standard deviation is $\sigma = 28.8675$. Our samples will again have $n = 80$; according to the Central Limit Theorem, the standard error of the mean in such samples will be $28.8675/\sqrt{80} = 3.28$. Thus, the Central Limit theorem predicts that the means all possible of 80-observation samples from this population will follow a normal distribution whose mean is 50 and standard error is 3.28. Let's see how well the theorem predicts the results of this simulated experiment. Given the approach of our simulation, we should find that the sample mean of column 81 will be about 50, and the standard deviation of the column will be a good approximation of the theoretical standard error of the mean.

🖱 **Calc ➤ Random Data ➤ Uniform…** Generate **60** rows and store them in **C1–C80**. Specify a **Lower endpoint** of **0** and an **Upper Endpoint** of **100**.

🖱 **Calc ➤ Row Statistics…** Once again, select **Mean** of **C1–C80**, and **Store the results** in **C81**.

🖱 **Stat ➤ Basic Statistics ➤ Display Descriptive Statistics…** Select **C80** and **C81** once again, creating the histograms as in the earlier simulation (the results of one simulation are shown below).

6. *Describe the center, shape, and spread of your two graphs. Remember that yours will look different from the ones shown here, or from your classmates.*

7. *Look in the Session Window to find the mean and standard deviation of the two columns. Do the mean and standard deviation of C81 (Mean) match the theoretical values predicted by the Central Limit Theorem?*

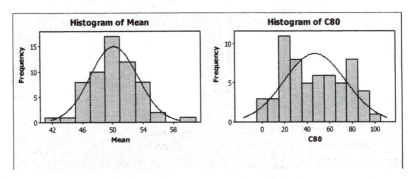

8. *What similarities do you see between your graphs and the two graphs shown here? What differences?*

9. *How do you explain the similarities and differences?*

## Sampling Distribution of the Proportion

The examples thus far have simulated samples of a quantitative random variable. Not all variables are quantitative. The Central Limit theorem and the concept of a sampling distribution also apply to qualitative random variables, with three differences. First, we are not concerned with the mean of the random variable, but with the *proportion* ($\pi$) of times that a particular outcome is observed. Second, we need to change our working definition of a "large sample." The standard guideline is that $n$ is "large" if both $n \cdot \pi > 5$ and $n(1-\pi) > 5$. Third, the formula for the standard error becomes:

$$\sigma_\pi = \sqrt{\frac{\pi(1-\pi)}{n}}$$

To illustrate, we'll generate more random data. Recall what you learned about Binomial experiments as a series of $n$ independent trials of a process generating success or failure with constant probability, $\pi$, of success. Such a process—two outcomes, with P(success) = $\pi$ each time— is known as a *Bernoulli trial*. We'll construct 60 more samples, each consisting of 80 Bernoulli trials:

🖰 **Calc ➤ Random Data ➤ Bernoulli...** Generate 60 rows in columns 1 through 80. Specify that **Probability of success** equals **0.3**.

This creates 80 columns of 0's and 1's, where "1" represents a success. By finding the mean of each row, we'll be calculating the relative frequency of successes in each of our simulated samples, also known as the *sample proportion, $\bar{p}$* .

🖰 **Calc ➤ Row Statistics...** Once again, select **Mean** of **C1–C80**, and **Store the results** in **C81**.

Now column 81 contains 60 sample proportions. According to the Central Limit Theorem, they should follow an approximate normal distribution with a mean of 0.3, and a standard error of

$$\sigma_\pi = \sqrt{\frac{\pi(1-\pi)}{n}} = \sqrt{\frac{(.3)(.7)}{50}} = .0648$$

As we have in each of the simulations, compute and graph the descriptive statistics for columns 80 and 81.

**10.** *Comment on the descriptive statistics and the graphs you see. How closely does the theorem describe your results?*

## Moving On...

The first questions continue to use simulated data, rather then a dataset.

1. What happens when $n$ is fewer than 30? Does the Central Limit Theorem work for small samples too? Using the Uniform distribution, try generating only 20 columns of data as you did above, and examining the histogram of the sample means. Describe the results of your experiment.

2. Try generating larger samples as well. Generate a greater number of columns and rows, and report on your findings.

3. This session used samples from two known distributions: one normal and one uniform. Minitab can generate random samples from a wide variety of other populations as well. Do a comparable demonstration using a uniform distribution ranging from –10 to 10, and report on the distribution of sample means from 80 samples of $n = 50$.

## Pennies

This file contains the results of 1,685 repeated in-class binomial experiments, each one of which consisted of flipping a penny 10 times. We can think of each 10-flip repetition as a sample of $n = 10$ flips; this file summarizes 1,685 different samples.

In all, nearly 17,000 individual coin flips are represented in the file. Each column represents a different possible number of heads in the ten-flip experiment, and each row contains the results of one student's repetitions of the 10-flip experiment. Obviously, the average number of heads should be 5, since the theoretical proportion is $\pi = 0.5$.

4. According to the formula for the standard error of the sample proportion, what should the standard error be in this case (use $n = 10$, $\pi = .5$)?

5. (Hint: For help with this question, refer to Session 8 for instructions on computing normal probabilities, or consult a normal probability table in your textbook). Assuming a normal distribution, with a mean $\mu = 0.5$ and a standard error equal to your answer to the previous question, what is the probability that a random sample of $n = 10$ flips will have a sample proportion of *0.25 or less*? (i.e., 2 or fewer heads)

6. Use the descriptive statistics commands to determine whether these real-world penny data refute or support the

predictions you made. How many of the samples contained 0, 1, or 2 heads respectively?

7. Comment on how well the Central Limit Theorem predicts the real-world results reported in your previous answer.

## Output

This file contains monthly time series data on industrial output in the United States for over 50 years beginning in January 1945. For these questions, we'll focus on one variable–**CapUtil**–which is that percentage of industrial capacity utilized in a given month. For illustrative purposes, we'll treat the observations in this file as a population, and we'll take several random samples from it to observe the variations in the sample means.

8. Use the descriptive statistics command to find the mean and standard deviation of **CapUtil**; we'll treat these as the population mean and standard deviation.

9. Use the Central Limit theorem to determine the standard error of the mean for random samples of size $n = 30$ drawn from this population.

10. As we did earlier in this session, use 30 empty columns of the worksheet to hold 50 rows of data randomly selected from the **CapUtil** column. To do this, you'll choose **Calc ➤ Random Data ➤ Sample from Columns...** and select **CapUtil** as the column from which to sample.

# Session 10

## Inference for a Population Mean

### *Objectives*

In this session, you will learn to:

- Interpret a confidence interval estimate
- Construct confidence intervals for a population mean
- Set up and perform a hypothesis test for a mean
- Interpret the results of a hypothesis test

### *Statistical Inferences from a Sample*

We often analyze sample data to go beyond the sample at hand and to draw a conclusion about the population from which we drew the sample. This process of *statistical inference* takes a variety of forms. In this chapter, we meet two of them.

You know that all random variables have center, shape, and spread. When we analyze a random variable we generally do not know its center, shape, or spread. In this session, we will concentrate on the *central location* of a distribution assuming that our only information about the variable comes from a random sample. We'll engage in the two essential inferential tasks:

- *Estimation*, where the goal is to determine the approximate location of the mean, and
- *Hypothesis testing*, where the goal is to decide if the mean is above or below a given value.

### *The Concept of a Confidence Interval Estimate*

A *confidence interval* is an estimate that reflects the uncertainty inherent in random sampling. In the prior session, we focused on the ways in which random sample results vary. To introduce the idea of a confidence interval, we'll start by simulating random sampling from a hypothetical normal population, with $\mu = 500$ and $\sigma = 100$. Unlike the prior session, though, we'll create only 20 samples, and place them into columns rather than rows.[1]

🖰 **Calc ➤ Random Data ➤ Normal...**   Generate 50 rows of data, store the results in **C1–C20**. Set $\mu = 500$ and $\sigma = 100$. This creates 20 samples of $n = 50$ observations.

🖰 **Stat ➤ Basic Statistics ➤ 1-Sample Z...**   This command will generate the 20 confidence intervals. In the dialog, select all 20 columns and specify that **Sigma** = 100. By default, the confidence level is 95%; we'll learn to change that soon.

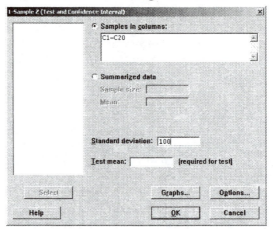

> 🖳 Remember that in a simulation, each of us will generate 20 different samples, and have 20 different confidence intervals. In 95% interval estimation, about 5% (1 in 20) of all possible intervals don't include $\mu$. Therefore, you may have 20 "good" intervals, or 19, or 18 or so.

Now maximize the Session Window and look at the output; the results of one simulation are shown on the next page. For each of the 20 columns, there is one line of output, containing the column number,

---

[1] This is due to the way the Confidence interval command operates; we can only construct intervals for data in columns.

variable name, sample size, mean, standard deviation, standard error, and a 95% confidence interval.

```
One-Sample Z: C1, C2, C3, C4, C5, C6, C7, C8, ...

The assumed standard deviation = 100

Variable   N     Mean     StDev   SE Mean        95% CI
C1        50   500.281   95.614   14.142   (472.563, 527.999)
C2        50   514.514   93.102   14.142   (486.796, 542.232)
C3        50   492.312   83.739   14.142   (464.594, 520.030)
C4        50   487.204   85.483   14.142   (459.485, 514.922)
C5        50   528.247   88.515   14.142   (500.529, 555.965)
C6        50   502.827  103.383   14.142   (475.109, 530.545)
C7        50   513.235   99.835   14.142   (485.517, 540.953)
C8        50   508.307   88.420   14.142   (480.589, 536.025)
C9        50   500.961   94.801   14.142   (473.242, 528.679)
C10       50   490.788  103.263   14.142   (463.070, 518.506)
C11       50   487.111  108.256   14.142   (459.393, 514.829)
C12       50   484.836  106.558   14.142   (457.118, 512.555)
C13       50   511.837  104.439   14.142   (484.119, 539.555)
C14       50   518.880   94.905   14.142   (491.162, 546.598)
C15       50   509.112  101.373   14.142   (481.394, 536.830)
C16       50   491.255  105.786   14.142   (463.537, 518.973)
C17       50   501.001  117.393   14.142   (473.283, 528.719)
C18       50   493.451  102.457   14.142   (465.733, 521.169)
C19       50   507.534  101.546   14.142   (479.816, 535.252)
C20       50   494.102   99.521   14.142   (466.384, 521.820)
```

In the sample output, one row is highlighted. In this simulation, the interval in this row lies entirely to the right of 500. Since this is a simulation, we know the true population mean ($\mu = 500$). Therefore, the confidence intervals ought to be in the neighborhood of 500, and indeed they are. However, due to the chance of random sampling, the fifth sample above happens to have had a particularly high sample mean, leading to this interval that fails to contain $\mu$.

1. *Do all of the intervals on your screen include 500? If some do not, how many don't?*

When we refer to a 95% confidence interval we are saying that 95% of *all* possible random samples from a population would lead to an interval containing $\mu$. Here you have generated merely 20 samples of the infinite number possible, but the pattern should become clear.

2. *If we were to repeat this exercise using 50 samples, about how many of the resulting 95% confidence intervals would include 500? Repeat this exercise using 50 samples; how many of your intervals included 500?*

## *Effect of Confidence Coefficient*

An important element of a confidence interval is the *Confidence Coefficient,* reflecting our degree of certainty about the estimate. By default, Minitab sets the confidence level at 95%, but we can change that value. These coefficients are conventionally set at levels of 90%, 95%, 98%, or 99%. Let's focus in on the impact of the confidence coefficient by re-constructing a series of intervals for the first sample column.

3.  ***What do you think will happen to the intervals if we change the confidence level to 90%?***

🖱 **Edit ➤ Edit Last Dialog…** We will re-do the intervals for all columns, but this time we'll change the **Confidence level** to 90%. In the 1-Sample Z dialog, click **Options**.

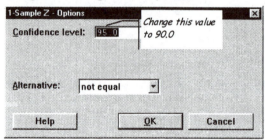

🖱 In the box marked **Confidence level**, change the **95.0** to **90.0**.

4.  ***Now how many intervals include 500? How do the 90% intervals compare to the 95% intervals?***

## *Dealing with Real Data*

Perhaps you now have a clearer understanding of what a confidence interval represents. It is time to leave simulations behind us, and enter the realm of real data where we generally don't know $\sigma$. For large samples (usually meaning $n > 30$), the traditional approach is to invoke the Central Limit Theorem, to estimate $\sigma$ using the sample standard deviation ($s$), and to construct an interval using the normal distribution, as the 1-Sample Z command does. In short, we find the sample standard deviation to estimate $\sigma$, and proceed as we just have. With such situations, it is better practice to use the **1-Sample t…** command instead; it automatically uses the sample standard deviation and builds an interval using the values of the *t distribution* rather than the normal.

Even with large samples, we should use the normal curve only when $\sigma$ is known. Otherwise the $t$ distribution is appropriate. In practice, the values of the normal and $t$ distributions become very close when $n$ exceeds 30. With small samples, though, we face different challenges.

## *Small Samples from a Normal Population*

If we cannot assume a population to be normal, we must take large samples or use non-parametric techniques such as those presented in Session 21. However, if we can assume that the parent population is normal, then small samples can be analyzed using the $t$ distribution. In Minitab, the procedural difference is quite minor, and involves choosing **1-Sample t...** rather than **1-Sample z...** on the **Stat** menu. To see how this works, we'll do one further simulation before moving on to a fully realistic situation. Let's take a small sample from a population which happens to be normal: voter turnout percentages in the state of Texas for the 2000 Presidential election.

🖰 **File ➤ Open Worksheet...** Select **Texas Votes**.

This worksheet contains the vote totals for each of the 254 counties in the state of Texas. Since our goal is to understand confidence interval estimation, let's treat this to as our population for the moment[2]. We will focus our attention on percent of eligible voters who cast votes for a Presidential candidate in November 2000. In the worksheet, this is the variable called **TurnOut**. Let's begin by finding the population mean ($\mu$) of **TurnOut**.

🖰 **Stat ➤ Basic Statistics ➤ Graphical Summary...** Select **TurnOut**.

Look at the Histogram with the superimposed Normal Curve. This strongly suggests that the underlying variable is normally distributed. From this output, we can also find the mean and standard deviation, along with other statistics.

To illustrate how we'd treat a small sample from a normal population, let's temporarily treat this column as a population, and select a small random sample from it. For the purpose of this illustration we will treat the mean of **TurnOut** (54.043%) as $\mu$.

---

[2] Our strategy is to sample from a known population to learn about how well samples can represent the entire population. Ordinarily, we only have sample evidence from which we draw conclusions about an unknown population.

🐭 **Calc ➤ Random Data ➤ Sample from Columns...** Sample 10 rows from **TurnOut**; store the sample in a new variable, **Small**.

🐭 **Stat ➤ Basic Statistics ➤ 1-Sample t...** Select **Small**, and look at the resulting interval.

5. *Does it contain the actual value of $\mu$? Will everyone in the class agree with you? Explain.*

## A More Realistic Case

This example illustrates how a confidence interval can usefully estimate a population mean. Ordinarily, we don't have population data; we only have our sample, and we use it to draw an inference about the population. Let's see how we would proceed with in a more realistic situation. Perhaps you have visited Yellowstone National Park in Wyoming and have seen an eruption of the Old Faithful geyser—so called because it reliably and predictably erupts at regular intervals. We'll construct a confidence interval for the mean time between eruptions.

🐭 **File ➤ Open Worksheet...** Open the file **Faithful**. This file contains sample data for 100 consecutive eruptions of the geyser.

We'll focus on the variable called **Wt1**, which represents the waiting time, measured in minutes, between eruptions during one five-day period. In this sample, $n = 100$ which is large enough to invoke the Central Limit Theorem. We can find a reliable confidence interval regardless of whether the population is normally distributed, but since we don't know $\sigma$, we still want to use the one-sample $t$ command.

🐭 **Stat ➤ Basic Statistics ➤ 1-Sample t...** Select **WT1**, and find the interval estimate in the Session Window.

```
One-Sample T: WT1

Variable    N     Mean    StDev  SE Mean        95% CI
WT1       100  71.6200  14.1527   1.4153  (68.8118, 74.4282)
```

The 95% confidence interval is between 68.81 and 74.43 minutes. What does the interval tell us? Since 95% of all random samples will yield a sample mean within approximately 2 standard errors from the population mean, it is a good bet that our sample is among the 95%. In other words, we could say that we are 95% confident that this interval contains $\mu$. Note that we are only making a statement about the likely value of the mean; this interval says nothing about the variation or shape of waiting times. It is a reasonable estimate of the *mean* waiting time.

⏳ Create a graphical summary of **Wt1**, as we did for **TurnOut** above.

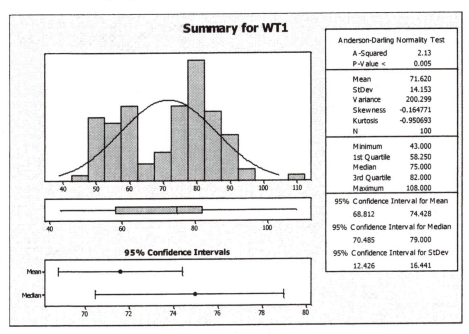

The graphical summary includes a 95% confidence interval for the mean. Notice that the observed waiting times ranged from 43 to 108 minutes, but that our confidence interval is much narrower than that—again, because we're estimating the *center* of the distribution.

In this example, we sought an estimate for the mean time between eruptions. Suppose, though, that you were visiting the park and had just missed an eruption. Suppose further that you needed to leave in about an hour and a quarter, and you wondered if Old Faithful would put on its display once more before you left. Such a case might stimulate a person to wonder about whether the mean waiting time is less than one hour or not? A question like this calls for a *hypothesis test*.

## The Logic of Hypothesis Testing

With confidence intervals, we wanted to estimate a population parameter. Sometimes, we are less concerned with finding an estimate as we are with comparing the parameter to a particular value. We now turn to this second kind of question. Although the questions are clearly related, they are different, and the methods for interpreting evidence are also different.

Random samples provide evidence about a population. We usually perform a hypothesis test when we have a hunch, a concern, or a hope about a parameter value; we represent that hunch in the *alternative hypothesis.*[3] To be cautious in weighing our sample evidence, we initially assume that our alternative—the belief we suspect, hope, or fear to be true—is false. We actually test the *null hypothesis*, which is the complement of the alternative[4]. For example, if we hope that Old Faithful will erupt, on average, in less than 75 minutes, we would state our hypotheses as follows:

| Hypothesis | Symbol | Expression |
|---|---|---|
| Null | $H_o$: | $\mu = 75$ |
| Alternative | $H_A$: | $\mu < 75$ |

The tests are set up to give substantial advantage to the initial belief, and only if the sample data are very compelling do we abandon our initial position. In short, the methods of hypothesis testing provide a working definition of *compelling evidence.*

In any test, we start with a *null hypothesis,* which is a statement concerning the value of a population parameter. We could, for example, express a null hypothesis like: "The mean weekly grocery bill for a family in our city is $150." The null hypothesis states a presumed value of the parameter. The purpose of the test is to decide whether data from a particular sample are so far at odds with that null hypothesis as to force us to reject it in favor of the alternative hypothesis.

## *An Artificial Example*

Before digging into real data, let's return to the first worksheet that we filled with simulated data (see page 126). Recall that this population was normal, with a mean of 500 and a standard deviation of 100. We will re-analyze the same data, this time testing to ask whether,

---

[3] See Utts, Jessica. *Seeing through statistics.2nd Ed.* (Pacific Grove CA: Brooks/Cole, 1999), Chapter 21.
[4] Some authors symbolically write the null as the complement of the alternative: if the alternative hypothesis were $\mu < 75$, the null would therefore be $\mu \geq 75$. Minitab always expresses the null as $\mu = 75$, so we'll follow the same convention in this book.

on the basis of each individual sample, we would conclude that $\mu = 500$ or not. Although we know $\sigma$, and therefore could use the 1-Sample Z command, let's imagine that we don't and use the 1-Sample t command.

🖱 **Stat ➤ Basic Statistics ➤ 1-Sample t...** Select all 20 columns, and specify that you want to **Test** (a) **mean** against a hypothetical value of 500.

Now look at Session Window (partial results of one sample shown on the next page). Remember that your output will be different.

The output reports the null and alternative hypotheses, and summarizes the results of these random samples. In this example, the first sample mean was 500.281. This is above 500, but then again we know in advance that half of all possible random samples have means above 500. Is this result so far from 500 as to cast serious doubt on the hypothesis that $\mu = 500$? The *test statistic* gives us a relative measure of the sample mean, so that we can judge how consistent it is with the null hypothesis. In this artificial test with a known population $\sigma$, the test statistic is computed as follows:

$$t = \frac{\bar{x} - \mu}{\sigma/\sqrt{n}} = \frac{500.281 - 500}{95.614/\sqrt{50}} = 0.02$$

```
 ┌──────────────────────────────────────────────────────────────────────┐
 │ One-Sample T: C1, C2, C3, C4, C5, C6, C7, C8, C9, C10, C11, C12, C13, C14, C15, │
 │                                              Your hypotheses            │
 │ Test of mu = 500 vs not = 500                                          │
 │                                                                         │
 │ Variable N     Mean    StDev  SE Mean        95% CI          T      P   │
 │ C1       50  500.281   95.614  13.522  (473.107, 527.454)   0.02  0.984 │
 │ C2       50  514.514   93.102  13.167  (488.054, 540.973)   1.10  0.276 │
 │ C3       50  492.312   83.739  11.842  (468.514, 516.111) -0.65  0.519  │
 │ C4       50  487.204   85.483  12.089  (462.909, 511.498) -1.06  0.295  │
 │ C5       50  528.247   88.515  12.518  (503.092, 553.403)   2.26  0.029 │
 │ C6       50  502.827  103.383  14.621  (473.446, 532.208)   0.19  0.847 │
 │ C7       50  513.235   99.835  14.119  (484.862, 541.608)   0.94  0.353 │
 │ C8       50  508.307   88.420  12.504  (483.178, 533.436)   0.66  0.510 │
 │ C9       50  500.961   94.801  13.407  (474.018, 527.903)   0.07  0.943 │
 │ C10      50  490.788  103.263  14.604  (461.441, 520.135) -0.63  0.531  │
 │  .          .        .       .            .               .      .      │
 │  .          .        .       .            .               .      .      │
 │  .          .        .       .            .               .      .      │
 └──────────────────────────────────────────────────────────────────────┘
```

This is the value reported in the *T* column of the output. In other words, 500.281 is 0.02 standard errors above the hypothesized value of $\mu$. Given what we know about approximate normal curves, that's not very far off. It is consistent with the kinds of random samples one would expect from a population with a mean value of 500. In fact, we could determine the likelihood of observing a sample mean more than 0.02 standard errors away from 500. That likelihood is called the *P-value* and it appears in the last column. In this instance, $P \le 0.984$; in other words, about 98% of all samples would give us such a result when $\mu$ really is 500. Notice also that the confidence interval for C1 does surround 500.

One way of thinking about the *P*-value is that if you were to reject the null hypothesis on the basis of this test, there is a probability of at most 0.984 that you are making a Type I error.[5] If you were unwilling to accept a 98% risk of a Type I error, you would be well advised against rejecting the null hypothesis in this instance.

In our confidence interval simulation, C5 contained an unusual sample. Look down the list of test statistics and *P*-values in the output on the prior page. For C5, the *T* statistic was 2.26 and the *P*-value was .029. For this sample, at a *significance level*[6] of $\alpha = 0.05$, we would *reject*

---

[5] Consult your textbook about Type I errors and about *P*-values.

[6] The *significance level* of a test is the complement of the confidence level, and is denoted by the Greek letter $\alpha$ (alpha). A significance level of 0.05 is equivalent to a 95% confidence level.

the null hypothesis, and erroneously conclude that the population mean is not equal to 500. This was the only sample in this simulation of twenty samples to have a misleading result.

Since this is a simulation, we know that the true population mean *is* 500. Consequently, we know that the null hypothesis really is true, and that most samples would reflect that fact. We also know that random sampling involves uncertainty, and that the population does have variation within it. Therefore, some samples will have sufficiently small *P*-values that we would actually reject the null hypothesis. In this simulation, one sample in 20 (5%) leads to this erroneous conclusion.

6.   *What happened in your simulation? Assuming a desired significance level of $\alpha$ = 0.05, would you reject the null hypothesis based on any of your samples?*

7.   *If you did reject the null hypothesis in one or more samples, did the accompanying confidence intervals include 500? What does this suggest to you?*

## A More Realistic Case: We Don't Know Sigma

Simulations are instructive, but are obviously artificial. This simulation is unrealistic in at least two respects—in real studies, we generally don't know $\mu$ or $\sigma$, and we have only one sample to work with. Let's see what happens in a more realistic case. Let's return to the **Faithful** worksheet.

Recall the hypothetical question we asked about waiting times between eruptions. We were hoping that Old Faithful has a mean waiting time of under 75 minutes, and therefore proposed these hypotheses:

$$H_o: \ \mu = 75$$
$$H_A: \ \mu < 75$$

The null hypothesis says that this sample comes from a population whose mean is 75 minutes. The *one-sided* alternative is that the sample was drawn from a population whose mean is less than 75.

This test has little practical importance, so we can tolerate a fairly high significance level ($\alpha$), such as 0.10. In other words, if we reject $H_o$, we'll settle for being 90% confident in our conclusion. Thus, the decision rule is this: We will reject the null hypothesis only if our test statistic has an associated *P*-value of 0.10 or less.

    **Stat ▸ Basic Statistics ▸ 1-Sample t...**   Select **WT1**, enter the hypothetical mean value of **75**. Because we want to change the significance level and choose a one-sided alternative, click on **Options**.

Change the confidence level to 90% and select a **less than** alternative, as shown.

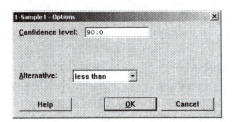

Now look in the Session Window at the results, shown below. Here the value of the test statistic, *t*, is –2.39 standard errors. That is to say, the sample mean of 71.62 minutes is relatively far from the hypothesized value of 75 minutes. The *P*-value of approximately 0.009 suggests that we should confidently *reject* null hypothesis, and conclude that the mean time between eruptions is less than 75 minutes (as we hoped it would be).

```
One-Sample T: WT1
Test of mu = 75 vs < 75
                                          90%
                                        Upper
Variable    N      Mean    StDev  SE Mean   Bound      T      P
WT1        100   71.6200  14.1527  1.4153  73.4459  -2.39  0.009
```

## *A Small-Sample Example*

In the previous example, we used a *t* test with a large sample. What happens when the sample is small? You may have learned that any *n* > 30 is "large," and that the Central Limit Theorem can apply in such cases. While that is a good guideline, the *t* test is preferable for samples in the neighborhood of 30 observations, especially since $\sigma$ is generally unknown in real-world data.

One bit of 'conventional wisdom' is that it is unhealthy to have a total cholesterol level in excess of 200. We have a data file with sample observations of 30 people who had recent heart attacks. One might suspect that their cholesterol levels would be above that level. Specifically, we'll consider their levels measured two weeks after the heart attack, using a variable called '14-day.' Let $\mu_{14\text{-day}}$ represent the mean of '14-day.' Formally, our hypotheses will be:

$$H_0: \mu_{14\text{-day}} = 200$$
$$H_A: \mu_{14\text{-day}} > 200$$

🖱 **File ➤ Open Worksheet...** Open **Cholest.**

Because this is a small sample, we may apply the $t$ distribution to a hypothesis test *only* if the underlying population is normally distributed (or at least roughly bell-shaped and symmetrical). We can check the assumption of normality by looking at the histogram of the sample data. If it is reasonably bell-shaped, we can proceed.

🖱 **Stat ➤ Basic Statistics ➤ Graphical Summary...** Select the variable **14-Day**.

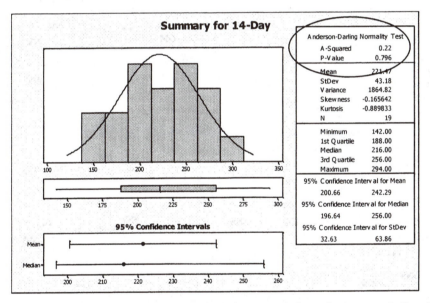

This command computes all of the usual descriptive statistics for the variable, displays a histogram of the data with a normal curve superimposed on the histogram, and also performs a statistical test of the null hypothesis that the sample was drawn from a normal population.[7] On the output, there is a reference to the *Anderson-Darling*

---

[7] This is one of a class of tests known as "Goodness of Fit" tests. Some of these procedures are presented more fully in Session 13.

*Normality Test*, with a test statistic called A-squared, and an associated *P*-value. When the *P*-value is small, we have sufficient evidence to reject the normality assumption. In this case, the *P*-value is quite high, suggesting that we may proceed with our *T* test.

🖰 **Stat ➤ Basic Statistics ➤ 1-Sample t...** In the dialog, select the variable **14-day**, specify a hypothesized value of 200. Now click **Options...**, and choose an alternative of greater than.

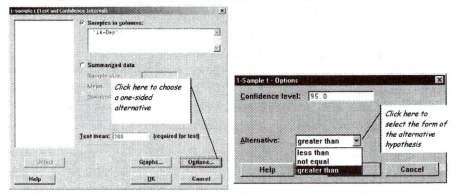

8.  **On the basis of this test, what is your conclusion? Is the mean cholesterol level above 200 among people with recent heart attacks?**

## Moving On...

### CPSVoting

This dataset contains results from the Current Population Survey. The questions refer to respondents' participation in the 1996 elections.

1.  Construct a 95% confidence interval for the mean age of registered voters in 1996. What does this interval tell you about the age of registered voters?

2.  Does this interval indicate that 95% of all registered voters fall in this age interval? Explain.

### Faithful

3.  Construct a 95% confidence interval for the mean waiting time between Old Faithful eruptions using the variable

**WT2** (a second sample of consecutive days). What does this interval tell us? How does this interval compare to the one we constructed earlier in the session?

4. Does this variable appear to be drawn from a normal population? What evidence would you consider to determine this?

## Cholest

5. The next two questions refer to the variables **2-day** and **Control**. Using the technique described in this session, determine whether it is reasonable to assume that each of these variables is normally distributed. Explain your reasoning.

6. Test the null hypothesis that the mean cholesterol level for the Control group of patients equals 200.

7. Test the null hypothesis that the mean cholesterol level for patients 4 days after a heart attack is less than or equal to 200.

## Swimmer2

This file contains the times for a team of High School swimmers in various events. Each student recorded two "heats" or trials in at least one event.

8. Construct a 90% confidence interval for the mean of first times in the 100-meter Free-Style.

9. Do the same for the second times in the 100-meter Free-Style.

10. Comment on the comparison of the two intervals you've just constructed. Suggest real-world reasons which might underlie the comparisons.

## Eximport

This file contains monthly data about the dollar value of U.S. exports and imports for the years 1948–1996.

11. Estimate the mean value of exports to the U.S., excluding military aid shipments. Use a confidence level of 95%.

12. Estimate the mean value of General Imports, also using a 95% confidence level.

13. On average, would you say that the United States tends to import more than it exports (excluding military aid shipments)? Explain, referring to your answers to #9 and #10.

14. Estimate the mean value of imported automobiles and parts for the period covered in this file, again using a 95% confidence level.

## Grades

This file contains cumulative grade point averages, SAT scores and high school rank data for 448 graduating college seniors.

15. Is it reasonable to assume that the Math SAT scores in this sample were drawn from a normally distributed population? Explain your thinking.

16. If we want to use this dataset for confidence interval estimation, how important is it to know if these Math SAT scores were drawn from a normal population? Explain.

17. Use a 98% confidence interval to estimate the mean Math SAT score for graduating students at this college.

18. Use a 95% confidence interval to estimate the mean cumulative GPA for graduating students at this college.

## JFKLAX

This is flight delay data for all flights between Kennedy International Airport (NY) and Los Angeles International Airport for a randomly selected date. Recall that the delay variables are measured in minutes; positive values indicate a late flight and negative values indicate an early departure or arrival.

19. Construct and interpret a 95% confidence interval for the mean departure delay.

20. Construct and interpret a 95% confidence interval for the mean arrival delay.

21. Which of these two intervals is wider? What facts explain why it is the wider of the two?

22. Does this set of data provide compelling evidence that, on average, flights depart late (i.e. more than 0 minutes late)?

23. Does this set of data provide compelling evidence that, on average, flights arrive late?

## LondonCO

These data were collected in West London and represent the hourly carbon monoxide (CO) concentration in the air (parts per million, or ppm) recorded between January 1 and June 30, 2000. For these questions, use the daily readings for the hour ending at 12 noon. You will perform one-sample $t$ tests with this variable.

24. In 1990, the first year of observations, West London had a mean CO concentration of 1.5 ppm. One reason for the routine monitoring was the desire to reduce CO levels in the air. Is there statistically significant evidence here that the CO concentration was reduced between 1990 and 2000? Explain.

25. Across town at London Bexley, the 2000 mean CO level was 0.4 ppm. Is there a significant difference between London Bexley and West London?

26. What about London Bridge Place, where the mean CO level was .3 ppm?

27. What do you think may account for the differences (or lack thereof)?

## Bowling

These are the scores for a 27-bowler amateur-bowling league for one randomly selected week. For each bowler, we have scores for each of three games ("strings") and the total score, which is simply the sum of the three strings.

28. Is it reasonable to assume that the variable called **STR1** was drawn from a normally distributed process? What evidence can you offer to support your answer?

29. Construct and interpret a 90% confidence interval for the mean score of first strings in this bowling league.

30. Construct and interpret a 90% confidence interval for the mean score of *third* strings in this league.

31. Suppose we wanted to construct a 90% confidence interval for the total score in this league. What problem do we encounter in creating a reliable interval estimate using this particular sample? How might we overcome the problem?

# Inference for a Population Proportion

## Objectives

In this session, you will learn to:

- Construct and interpret confidence intervals for a population proportion
- Perform and interpret hypothesis tests about a population proportion
- Make inferences about a population proportion using raw categorical data or summary data

## Inferences for Qualitative Data

In the prior session, we worked extensively with inferential techniques for the mean of a quantitative random variable. If we are interested in qualitative data, we can conduct similar inferential analysis. Of course, we cannot find the mean of a categorical variable, but we can think about the *proportion $\pi$* of the population that falls into one particular category. Such instances are the topic of this session.

## Confidence Interval for a Population Proportion

In one of the previous examples, we looked at the percentage of registered voters who actually cast ballots in the 2000 Presidential election. In that example, our data were aggregated by county for the state of Texas. Let's look at another election-related dataset, this one containing survey responses given by registered voters nationwide for the 1996 Presidential election.

⌨ **File ➤ Open Worksheet...** Open the file **CPSVoting**.

The Bureau of Labor Statistics and the Bureau of the Census jointly administer the Current Population Survey (CPS), interviewing large numbers of residents selected by random sampling. Our worksheet contains responses to ten questions posed shortly following the election on November 5, 1996; in this dataset, all respondents are randomly selected from among registered voters replying to the CPS. Thus, they are all eighteen years of age or older, are all U.S. citizens, and are all registered to vote. Let's use this sample to estimate voter turnout nationwide in the 1996 election. The relevant variable in this instance is **Voted**, which is coded "Yes" if the respondent answered "yes" to this question:

> *In any election some people are not able to vote because they are sick or busy or have some other reason, and others did not want to vote. Did (you/ name) vote in the election held on Tuesday, November 5?*[1]

🖐 **Stat ➤ Basic Statistics ➤ 1 Proportion...** Just select the variable **Voted**, and click **OK**.

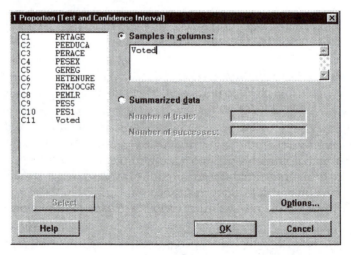

The interval estimate will appear in the Session Window:

---

[1] Also see the variable called **PES1** which is the respondent's direct answer to the question. Some people could not recall or refused to answer, and thus there are more than two possible answers to the question. For this illustration, the answers have been reduced to Yes and No.

```
Test and CI for One Proportion: Voted
Test of p = 0.5 vs p not = 0.5

Event = Yes
                                                    Exact
Variable     X     N  Sample p        95.0% CI    P-Value
Voted      183   387  0.472868  (0.422208, 0.523946)  0.309
```

Note that this command does double duty. It computes a confidence interval and performs a hypothesis test for these data. We'll consider hypothesis tests shortly, but let us consider this confidence interval first.

Note also that the output reports that "Event = Yes" indicating that Minitab has computed the sample proportion of "Yes" responses (i.e. 183 of 387 respondents said Yes). Here the 95% confidence interval is from .422 to .524, indicating that we are 95% confident that approximately 42% to 52% of all registered voters cast ballots in the 1996 Presidential election. If we wanted to use a different confidence level, we can do so by clicking the **Options** button and entering the value we want.

Now consider the hypothesis test results.

1. *Look at the Session Output. What are the null and alternative hypotheses in this test?*

2. *Based on the reported P-value, what should you conclude about the value of $\pi$?*

By default, Minitab has selected a common hypothesized value for the population proportion $\pi$. Recall that in Session 10 we estimated that about 54% of Texans voted in the 2000 Presidential Election. Suppose we wanted to know if the same proportion voted nationally in 1996. Then our hypotheses would be:

$$H_o: \pi = 0.54$$
$$H_A: \pi \neq 0.54$$

To test this null hypothesis, we return to the prior dialog, and this time click on **Options** (see dialog on next page). Enter the hypothetical value of 0.54, and then click **OK**.

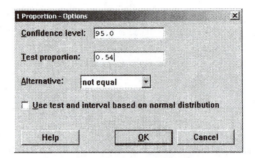

Now run the test and look at the output. The confidence interval is unchanged since we are still estimating the population proportion at a 95% confidence level using the same data. But look at the hypothesis test results: the *P*-value is only .009, indicating that we should reject the null hypothesis, and conclude that the proportion of people who voted in 1996 was not 0.54.

**3.    *The confidence interval lies entirely below 0.54. Do you think it is possible to run a test in which we find a 95% confidence interval that includes the hypothesized value, but in which we reject the null at a 5% significance level? Explain your thinking.***

## A One-Sided Confidence Interval

Suppose we had asked whether the population proportion was *less than* 0.54, rather than asking the two-sided question of whether the proportion was not equal to 0.54. As you might surmise, we would have specified a **less than** alternative in the previous dialog, and evaluated the results. Let's do just that.

🖱  Re-run the last test, this time choosing **less than** for the alternative hypothesis. You'll see this output:

**Test and CI for One Proportion: Voted**
```
Test of p = 0.54 vs p < 0.54

Event = Yes
                                                      Exact
Variable     X     N   Sample p   95.0% Upper Bound  P-Value
Voted      183   387   0.472868             0.515953  0.005
```

Here, you see that we don't get a confidence interval, but instead see only the *95% Upper Bound* value of 0.51593. What happened to the interval?

Since we are performing a one-sided test, we get a one-sided interval. More to the point, because we are asking if this sample gives us sufficiently compelling evidence to conclude that $\pi$ is less than 0.54, the relevant estimation question is "What is the highest plausible value of $\pi$ at a 95% confidence level"? The resulting upper bound can be interpreted as being the highest plausible value of the population parameter based on this particular sample and a 95% confidence level.

We should also note that Minitab reports one-sided confidence levels for means as well as proportions. Given the breadth of Session 10, we did not specifically explore the one-sided intervals at that time.

## Variables with More than Two Categories

The variable called **Voted** had only two possible values, Yes and No. Of course, many categorical variables have multiple possible values. If we were interested in the population proportions of all possible categories, we'd need to use a different procedure altogether. You can find those details in Session 13.

Even with a multi-category variable, we might want to focus on the proportion of responses that fall into a single category. In such a case, we will define a new variable with just two possible values, one that reflects the category of interest, and one that comprises all others. Let's see one simple way to handle such a situation.[2]

## Inference Using Summary Data

Consider the signs of the Zodiac. Since there are twelve birth signs, it reasonable to believe that each sign should be reported by approximately one in twelve respondents if we were to ask people about their birth sign. Suppose that we wanted to know if one-twelfth of all people reported having Aries as their sign.

Let's look at some data from the General Social Survey, a biennial project that gathers information about social attitudes and behaviors in the United States. Among other important questions, we find that GSS researchers ask about the Zodiac signs. Open the worksheet called **GSSGeneral**.

One simple way to attack this problem is to see how many people in the sample reported Aries, and then compare that sample proportion to the hypothesized value of 1/12 (0.08333).

---

[2] As is often the case, there is more than one way we could do this. For the sake of simplicity, we'll see one approach.

🖱 **Stat ➤ Tables ➤ Tally...** Select the variable **Zodiac**, and construct the simple tally.

You should find that 109 people out of 1423 said that their sign is Aries.

🖱 **Stat ➤ Basic Statistics ➤ 1-Sample Proportion...** Now have summary data, rather than data in one column (remember that **Zodiac** has 12 possible values rather than just two). Complete the dialog as shown; under **Options**, test the proportion .0833 against a **not equal** alternative.

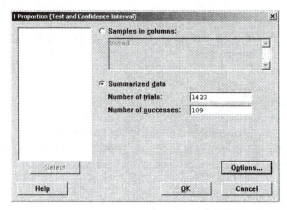

In the output, the confidence interval is (0.063315, 0.091661) and the *P*-value is 0.363. Based on this result, we can't reject the null hypothesis. In other words, there is insufficient evidence here to discredit the idea that one-twelfth of the population was born under Aries.

## Moving On...

Now apply what you have learned in this session to each of these questions. Explain what you conclude from each test, and why. Unless otherwise noted, perform tests at a significance level of $\alpha = 0.05$.

### CPSVoting

1. Use a 90% confidence interval to estimate the proportion of registered voters who are males. Based on this evidence, would you say that about half of the respondents are men? Explain your thinking.

2. Based on this evidence, would you conclude that more than 3% of the population was unemployed (either looking or on layoff)? Explain your thinking.

## Student

3. It is generally estimated that left-handed people make up about 10% of the U.S. population. Do these students appear to be drawn from such a population?

4. Several years ago, about 30% of my students reported having smoked at least one cigarette in the past month. Based on the evidence in this sample, has that proportion increased recently?

5. Construct 95% confidence interval estimates for the proportion of these students who:

   - Own a dog
   - Own a car
   - Know someone who has been struck by lighting

6. Do any of these intervals overlap? What does that indicate?

## GSSGeneral

7. In the Session, we asked if one-twelfth of the population was born under Aries. Perform a similar test corresponding to the sign Libra. What do you conclude?

8. Do the same for Capricorn. What do you conclude?

## JFKLAX

9. The variable called **LateArr** is coded as **Yes** for a flight that arrived late, and **No** for a flight that was not late. Does this data set provide compelling evidence that more than one-fourth of flights arrive late?

10. Assuming these data represent a typical day, do more than half of the flights between New York and Los Angeles depart from L.A.?

11. Does American Airlines run approximately half of the flights between New York and Los Angeles? Explain.

## GSSSex1

12. Use the variable called **Stray** to decide if more than 10% of adults have ever engaged in extramarital sex while married.

13. Public health specialists recommend the use of condoms, particularly for unmarried people. This dataset has two variables related to condom use. In both instances, respondents were asked if they or their partner used a condom the last time they had sexual intercourse. The variable **CONDOM** includes all respondents; the variable **CondNoMarr** includes only unmarried respondents. Before looking at the data, speculate about what percent of respondents reported having used a condom. Use your guess as the hypothesized proportion, and perform the appropriate test. Report your results, and comment on what you find.

## Violations

This worksheet contains records of motor vehicle violations on a college campus for one particular semester.

14. Estimate the proportion of tickets written to Freshmen. If we assume that first-year students make up 25% of the college community, would you judge that first-year students less than their "fair share" of tickets?

15. A Student Affairs official was quoted in the student newspaper as saying "About 75% of the tickets written are for Parking Lot violations." Does this sample cast significant doubt on that statement?

16. In the past, 80% of all fines were for $10. Does this sample provide compelling evidence that the proportion has risen?

## Web

This data set contains the results of an experiment using the Random Yahoo© Link, a feature of Yahoo.com. The random link literally will take the web surfer to a website randomly selected from Yahoo's massive database.

17. At any given moment, a user may encounter a problem connecting to a given website. A researcher clicked on the random link 20 times, recording the number of problems encountered. He repeated this process for a total of 20 samples (i.e. 400 attempted connections in all). Use the **Calc ➤ Column Statistics...** command to determine the total number of problems encountered. How many problems did he have in his experiment?

18. Provide a 95% confidence interval estimate for the proportion of web sites that might present connection problems. Comment on what you find.

19. Prior to the experiment, the research predicted he would find that approximately 3% of all web sites would have connection problems. Based on this sample, should he reject that prediction? Discuss your findings.

# Inference for Two Samples

## *Objectives*

In this session, you will learn to:

- Draw inferences about the difference between two population means
- Draw inferences about the difference in means of two "matched" samples drawn from a population
- Draw inferences about the difference between two population proportions

## *Working with Two Samples*

In the two previous sessions, we learned to analyze a single sample and make inferences about a population. Often the goal of a study is to *compare* an attribute in two distinct populations. To make such comparisons, we must select two independent samples, one from each population.

What does it mean to say that two samples are independent? We must construct the samples so as to insure that the observed values in one sample could neither affect nor be affected by the observations in the other, and that the two sets of observations do not arise from some shared factor or influence.

We'll begin by comparing the means of two populations. We know enough about random sampling to predict that any two samples will likely have different sample means *even if they were drawn from the same population.* We typically anticipate some variation between any two sample means. Therefore, the mere fact that we find some discrepancy between two sample means is not compelling evidence. In comparing two samples, we ask: Is the observed difference between two sample means extraordinarily large? Specifically, is it large enough to convince us that the populations have different means?

Our first illustration draws on some historic work in physics—some early measurements of the speed of light. In the second half of the 19th century, physicists in Europe and the United States worked on devising methods to measure the speed of light accurately. In 1879, A.A. Michelson took a series of measurements of the velocity of light in air over a distance of 600 meters, refining techniques earlier developed by Foucault. In 1882, he used a modified methodology, and recorded additional readings. We'll want to determine if he obtained significantly different results with the second, more accurate approach. Open the **Michelson** worksheet, which contains two columns. Column C1 contains the measurements of light speed, recorded in kilometers per second (km/sec). Column C2 contains the year the measurements were collected.

We can restate our question in formal terms as follows:

$$H_o: \mu_{1882} - \mu_{1879} = 0$$
$$H_A: \mu_{1882} - \mu_{1879} \neq 0$$

Note that the hypotheses are expressed in terms of the *difference* between the means of the two groups. The null hypothesis says that the mean measurements are the same for the two years, and the alternative is that they are different.

You may have learned that a 2-sample $t$ test requires three conditions:

- Independent samples
- Normal populations
- Equal population variances (for small samples)

The last item is not actually required to perform a $t$ test. It is true that we compute the test statistic differently when the variances are unequal, but Minitab can readily handle situations where we do not assume equal variances. When population variances are unequal, treating them as equal may lead to a seriously flawed result. You'll see that the distinction is minor in terms of using the software. Before checking for equality of variance, we should decide if these data come from normal populations.

In practice, the $t$ test is generally reliable as long as the samples suggest symmetric, bell-shaped data without substantial departures from a normal distribution. Thus, we are well advised to examine our data for normality.

⌐🖰 **Graph ➤ Probability Plot ...** We want a single plot of **Speed**.

 ⌐🖰 Click on **Multiple Graphs...** In the **Multiple Variables** tab, choose In separate panels on the same page.

 ⌐🖰 Then click the **By Variables** tab, and select the variable **Year** under By variables with groups in separate panels. Click OK in both dialogs.

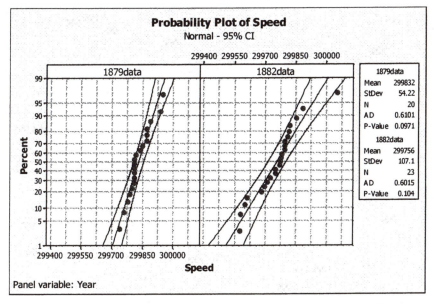

In these graphs, we see some points that depart from the 45-degree line. The Anderson-Darling (AD) Normality Test, with its null hypothesis of normality, has a $P$-value of approximately 0.10 for both sets of observations. This casts some doubt on the assumption of normality, but we will proceed for the sake of this first illustration.[1]

Minitab provides two formal tests for the equality of two variances. We choose the appropriate test statistic depending on whether or not we have normal data. In both tests, the null hypothesis is that the

_____

[1] Session 21, on nonparametric techniques, addresses the situation in which we have non-normal data. Minitab also provides other methods for checking normality. The interested reader should try **Stat ➤ Basic Statistics ➤ Normality test....**

variances are equal; if the *P* value is less than our significance level (α), we reject the null and determine that the variances are unequal.

> 🖱 **Stat ➤ Basic Statistics ➤ 2 Variances…** Since all of the speed data is in one column, we choose **Samples in one column**, select **Speed** for **Samples** and **Year** for **Subscripts**.[2] Click **OK**.

This command generates both a graph (shown below) and Session Window output. When we have normal data, we use the *F*-test; for other continuous data, we rely on Levene's test. For present purposes, let's assume an α value of 0.10. In this case, both tests present sufficient evidence that we should treat the variances as unequal. Visually, note the spread of the boxplots.

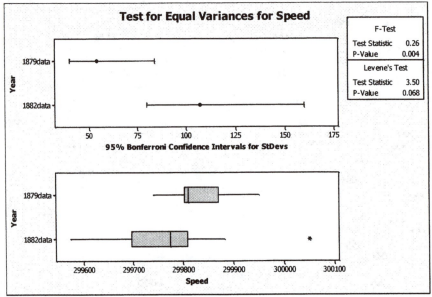

Now we are ready to conduct the *t* test.

> 🖱 **Stat ➤ Basic Statistics ➤ 2-Sample t…** As in the prior dialog, we choose **Samples in one column**, select **Speed** for **Samples** and **Year** for **Subscripts**.

---

[2] Minitab refers to this data arranged this way as "stacked." In most of our examples, we'll use stacked data. See Appendix C for a full discussion of the reasons for stacking or unstacking data.

🖰 Before clicking **OK,** click on the **Graphs** button, and choose **Boxplots.** Click **OK** in the **2-Sample t-Graphs** dialog.

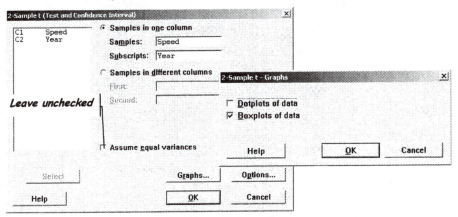

🖰 Leave the **Assume equal variances** box unchecked, and click **OK**. You'll see this graph:

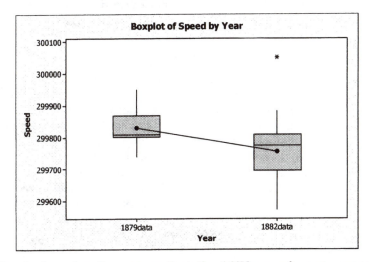

In the boxplot, it appears that the 1879 speeds were generally higher than the 1882 measurements. The dots near the median line in each box represent the sample means. The test results in the Session Window will tell us whether the difference in sample means is so large as to reject the null hypothesis in favor of the alternative.

---

**Two-Sample T-Test and CI: Speed, Year**

```
Two-sample T for Speed
                              SE
Year       N    Mean  StDev  Mean
1879data  20  299831.5  54.2   12
1882data  23   299756   107    22


Difference = mu (1879data) - mu (1882data)
Estimate for difference:  75.2826
95% CI for difference:  (23.5788, 126.9864)
T-Test of difference = 0(vs not =): T-Value = 2.96 P-Value = 0.006 DF = 33
```

---

We interpret the output much in the same way as in the one-sample test. The test statistic $t$ equals 2.96. Since the $P$-value here is so small, we reject the null hypothesis and conclude that the mean speed measurements in 1879 and 1882 were indeed different.

1. **How should we interpret the 95% confidence interval estimate for the difference in means?**

2. **Re-run this test, checking the box marked "Assume Equal Variances." How does the resulting output differ from the output shown above?**

## Matched vs. Independent Samples

In the prior example, we have focused on differences inferred from two independent samples. Sometimes, though, our concern is with the *change* in a single variable observed at two points in time. For example, to evaluate the effectiveness of a weight-loss clinic with 50 clients, we need to assess the change experienced by each individual, and not merely the collective gain or loss of the whole group.

Such a study design is sometimes called "matched samples," "repeated measures," or "paired observations."[3] Since the subjects in the samples are the same, we pair the observations for each subject in the sample, and focus on the difference between two successive observations or measurements. We'll start with another example of historic importance.

🖰 Open the worksheet called **Gosset**.

---

[3] Matched samples are not restricted to "before and after" studies. Your text will provide other instances of their use. The goal here is merely to illustrate the technique in one common setting.

William S. Gosset, writing under the pen name *Student,* published a 1908 paper detailing the *t* distribution. Our worksheet contains some of the data appearing in that paper. The measurements refer to an experiment in which eleven different varieties of barley seed were sown.[4] Half of the seeds were kiln-dried before planting—a practice then believed to increase crop yield (pounds per acre). The other seeds were not dried prior to planting. To control for variations in soil, light and water conditions, the seeds were planted in eleven adjacent pairs of "split" plots. In that sense, each pair of measurements in the worksheet must be analyzed as a pair, because they each share important characteristics.

We are performing this test to see if there is compelling evidence that kiln-drying increases crop yield. So, in this test our null hypothesis is to the contrary, which is that there is no improvement in yield between the two kinds of seed. Let $D_i$ represent the difference in yield in split plot *i*, and $\mu_D$ equal the mean of $D$.

$H_o$: $\mu_D = 0$  {there is no difference in yield}
$H_A$: $\mu_D > 0$  {there is a difference in yield}

Because the plants grown in a single plot shared the same soil, light, and water conditions our samples are not independent. As such, we use a different procedure to conduct the test.

🖱 **Stat ➤ Basic Statistics ➤ Paired t...** In the dialog, select **KD** as the first variable, and **NKD** as the second.

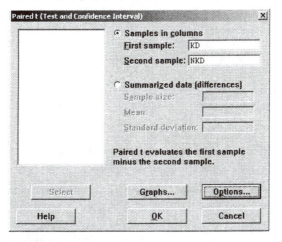

─────────────────────────

[4] Gosset attributes the data to Dr. Voelcker, whose results were reported in the *Journal of the Agricultural Society.*

🖱 Create a **Boxplot of differences** by clicking on **Graphs**.

🖱 In the **Options** subdialog, select a **greater than** alternative hypothesis.

Here is the resulting output.

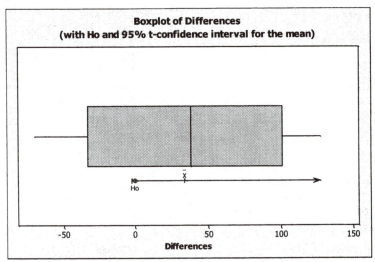

**Boxplot of Differences**
**(with Ho and 95% t-confidence interval for the mean)**

Differences

```
Paired T-Test and CI: KD, NKD
Paired T for KD - NKD

               N     Mean     StDev    SE Mean
KD            11   1875.18   332.85    100.36
NKD           11   1841.45   342.74    103.34
Difference    11   33.7273   66.1711   19.9513

95% lower bound for mean difference: -2.4338
T-Test of mean difference = 0 (vs > 0): T-Value = 1.69   P-Value = 0.061
```

> 3. *What do you conclude about the impact of using kiln-dried seed? (Assume a 0.05 significance level)*

> 4. *In the boxplot, what does the point marked $H_0$ represent? What do the red line and the point marked $\bar{x}$ represent?*

As noted earlier, we should correctly treat this as a paired sample $t$ test. But what would happen if we were to (mistakenly) treat the data as two *independent* samples? Let's try it and see.

🖰 **Stat ➤ Basic Statistics ➤ 2-sample t...** Here we have data in two columns (**NKD, KD**), and our alternative hypothesis is that the KD population mean is greater than the NKD population mean. Run the test assuming equal variances.

5.   *How does this result compare to prior one?* In the correct version of this test, we saw a fairly large, but not statistically significant difference—the *t*-statistic was 1.69, significant at the 0.10 level.

6.   *Why does this test have a different result? Does kiln-drying increase yields or does it not?*

This example illustrates the importance of knowing which test applies in a particular case. Minitab is readily able to perform the computations either way, but the onus is on the analyst to know which method is the appropriate one. As you can see, getting it "right" makes a difference—a substantial effect "disappears" when viewed through the lens of an inappropriate test.

## Comparing Two Proportions

Suppose we want to know whether male college students are more likely than female students to own a car. After all, one common image of the American male is that he is particularly devoted to his automobile. Let's return to the Student data to see if it can shed light on the question. Open the **Student** worksheet.

This question involves a binary categorical variable, car ownership (**Owncar**) and we'll want to compare the proportion of owners across **Gender**. These are stacked data, with all students' responses to the car ownership question in one column, and subscripts identifying gender in another. Because the first student in the sample is a female, Minitab will compute the difference in proportions as $\pi_{female} - \pi_{male}$.

🖰 **Stat ➤ Basic Statistics ➤ 2 Proportions...** As shown in the completed dialog box, **OwnCar** is the **Sample** and **Gender** contains the **Subscripts**.

🖰 Click **Options** to specify a **less than** alternative hypothesis.

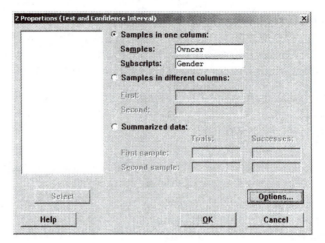

Before looking at the results, there is one restriction in the use of this command: The sample may not contain missing data, or the command will not execute. If there are missing observations, subset the worksheet, omitting rows where that variable is missing data. Alternatively, use a cross-tabulation to summarize the number of 'successes' for each sample, and then simply enter the figures in the dialog under the **Summarized data** option.

The results are these:

---

**Test and CI for Two Proportions: Owncar, Gender**

```
Success = Yes

Gender    X    N   Sample p
Female   44   52   0.846154
Male     47   59   0.796610

Difference = p (Female) - p (Male)
Estimate for difference:  0.0495437
95% upper bound for difference:  0.168720
Test for difference = 0 (vs < 0):  Z = 0.68   P-Value = 0.753
```

---

This command computes the sample proportions for each sample, along with point and interval estimates for the difference between the two population proportions. As for the test, the test statistic follows an approximate normal distribution;[5] in this case, the *P*-value is 0.753, which leads us to conclude that we fail to reject the null. These samples provide no convincing evidence that male students own cars in greater proportions than females. Indeed, in this sample, the women were slightly more likely to be car owners than the males.

## Further Reading

Gosset, W.S. (1908) The probable error of a mean. *Biometrika* 6, pp. 1–25.

Stichler, R.D., Richey, G.G., and Mandel, J. (1953). Measurement of treadwear of commercial tires. *Rubber Age* 73:2.

Stigler, S.M. (1977) Do robust estimators work with real data? *Annals of Statistics* 5(6), 1055–1098.

## Moving On...

Use the techniques presented in this lab to answer the following research questions. Justify your conclusions by citing appropriate test statistics and *P*-values. Unless otherwise noted, use $\alpha = 0.05$.

### Student

1. Do commuters and residents earn significantly different mean grades?

2. Do car owners have significantly fewer accidents, on average, than non-owners? (Hint: Look closely at Session output.)

3. Do dog owners have fewer siblings on average than non-owners?

4. Many students have part-time jobs while in school. Is there a significant difference in the mean number of hours of work for males and females who have such jobs? (Omit students who do not have outside hours of work by subsetting the worksheet.)

5. Are males as likely as females to be dog owners?

---

[5] The test statistic distribution is approximately normal for large enough samples. For each of the samples, if $n\pi$ and $n(1-\pi)$ both exceed 5, we have sufficient data.

## Gosset

6. The third column of the worksheet is a variable called **Diff**. Perform a 1-sample *t* test of the null hypothesis that the mean of this variable is less than or equal to 0. Compare the results to the example shown in this session, and comment on your findings.

## Faithful

This dataset contains measurements of the minutes elapsed between eruptions of the Old Faithful geyser. Each column represents 100 measurements documented during two different periods in August 1985.

7. Does each of these variables appear to come from a normally-distributed parent population?

8. Before looking at the sample variances, comment on why we might expect the variances of the two sets of measurements to be similar.

9. Perform an appropriate test to determine whether the data suggest a significant difference between the two population variances.

10. Estimate the difference between the means; is it reasonable to conclude that there was a statistically significant change in Old Faithful's eruption pattern between the two periods?

## GSSEduc

These are responses of 1445 U.S. adults to the 1998 General Social Survey. All of the variables relate to education.

11. Does this sample provide statistically significant evidence that in 1998 U.S. adults had completed more years of school than their fathers had? Explain.

12. Does this sample provide statistically significant evidence that in 1998 men had completed more years of school than women? (refer to the variables **EDUC** and **SEX**). Explain.

### GSSSex1

These are responses of 1445 U.S. adults to the 1998 General Social Survey. All of the variables relate to attitudes toward sex and marriage.

13. Respondents were asked if they had ever strayed during marriage. The variable **STRAY** represents the answers given by those who had were or had been married; a value of 1 represents 'yes.' Is there statistically significant evidence that married men stray more often than married women?

14. **Condom** is a variable representing responses to the question "Did you use a condom the last time you had sex?" Were men more likely to have said "yes" than women?

### Swimmer1 *(Note: These are challenging problems, requiring manipulation of the worksheet before analysis)*

15. Do individual swimmers significantly improve their performance between the first and second recorded times?

    NOTE: To answer this question using the techniques of this session, you must "unstack" the **Time** data, as follows:

    🖱 **Data ➤ Unstack Columns...** Complete the dialog as shown. Do a paired-sample test using the two new columns.

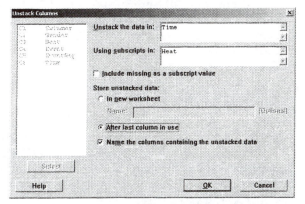

16. In their second times, do swimmers who compete in the 50-meter Free-Style swim faster than those who compete in the 50-meter breaststroke? (Subset the worksheet.)

17. In the first 50-meter Free-Style heats, do the men swim faster than the women? (Subset the worksheet.)

## CPS Voting

This worksheet contains a sample of responses from the Current Population Survey of 1998. All of the responses in this dataset relate to voting behavior.

18. Did men and women in the United States vote in equal proportions in the 1996 general election? Explain your reasoning, referring to results of your statistical analysis.

19. Among people who described themselves as Employed—At Work (see variable **PEMLR**), did men and women vote in equal proportions in the 1996 general election? Explain. [Hint: subset the worksheet, including only those rows where **PEMLR**=Employed—At Work).

## StateTrans

This worksheet contains data related to transportation in the United States and Washington D.C. in 1998.

20. All of the states set a legal limit for blood alcohol concentrations while operating a motor vehicle. Twelve states use a limit of 0.08 and 39 use a limit of 0.10. Assuming that 1998 is a representative year, is there compelling statistical evidence that drivers in the 0.08 jurisdictions travel more miles per capita annually than those in the 0.10 jurisdictions? Because one of the samples is small, comment on the assumptions of normality in this case. Also, comment on the equal variance assumption.

21. Assuming that 1998 is a representative year, is there statistically significant evidence in this dataset that the 0.08 jurisdictions have more traffic fatalities than the other states? Because one of the samples is small, comment on the assumptions of normality in this case. Also, comment on the equal variance assumption.

## Haircut

This worksheet holds responses from a sample of 60 college students. Students were asked "How much did you pay for your last professional haircut?" These questions compare the responses of the male and female students.

22. Create side-by-side boxplots of the data for men and women. Is it reasonable to think these two populations share the same variance? Explain.

23. Is there a statistically significant difference in the prices paid by men and women for haircuts?

## Tirewear

This worksheet represents a sample of 16 tires, each of which was measured using two different methods to estimate the amount of wear. The first method estimates wear as a function of the difference between the original weight of the tire and the weight at the time of measurement. The second method estimates wear based on the depth of the grooves, or treads, on the tire. All recorded measurements are in thousands of miles.

24. Let's assume that the goal of the study was to compare the results of these two methods of measurement. As a matter of experimental design, why was it wise to measure each tire using two methods, rather than measuring one sample by the weight method and an independent sample by the groove method.

25. Is there a statistically significant difference in wear estimates using these two methods? Explain your conclusion, referring to the results of an appropriate test.

## Water

26. Is there statistically significant evidence here that water resources sub-regions were able to reduce irrigation conveyance losses (i.e., leaks) between 1985 and 1990?

27. Did mean per capita water use change significantly between 1985 and 1990?

## LondonCO

This file contains carbon monoxide (CO) measurements in West London air by hour for the year 2000. Measurements are in parts per million.

28. Would you expect to find higher CO concentrations at 9 AM or 5 PM? Why? What might account for different levels of CO at different times of the day? Perform an appropriate $t$ test and comment on what you find. Be sure to decide if the variances should be assumed equal or unequal, and explain how you made your judgment.

29. Would you expect to find higher CO concentrations at 7 AM or 9 PM? Why? What might account for different levels of CO at different times of the day? Perform an appropriate $t$ test and comment on what you find. Be sure to decide if the variances should be assumed equal or unequal, and explain how you made your judgment.

## Cholest

30. Comparing the heart attack patients on Day 2 to the control patients, we might expect the latter group to have lower cholesterol readings. Do the data support that conclusion at a 0.05 significance level?

31. Is there a significant reduction in cholesterol levels for the patients between Day 2 and Day 14?

## SanDiego Crime

These figures represent the number of reported crimes in the neighborhoods of San Diego California during the month of November 2000.

32. Assuming that November 2000 was a typical month, is there statistically significant evidence to suggest that there are more Residential burglaries than Commercial burglaries in San Diego? Explain how you arrived at your conclusion.

33. Comment on the extent to which the assumptions for your statistical test seem to be satisfied by this dataset.

34. Again assuming that November 2000 was a representative month, is there significant evidence that property crimes (**Total Property**) are more prevalent than violent crimes (**Total Violent**)? Before drawing your conclusion, comment on the degree to which the assumptions for your test are met by this dataset.

# Chi-Square Tests

## *Objectives*

In this session, you will learn to:
- Perform and interpret Chi-square[1] goodness of fit tests
- Perform and interpret Chi-square tests of independence

## *Review of Qualitative vs. Quantitative Data*

All of the tests we have studied thus far have been appropriate exclusively for quantitative variables. The tests presented in this session are suited to analyzing *qualitative* or discrete quantitative variables, and the relationships between two such variables. The tests fall into two categories: Goodness of fit tests and tests for independence.

## *Goodness of Fit Testing*

When we construct a mathematical model of a process or phenomenon, we often begin with a sample, and develop a model that reliably describes or simulates the empirical data. Then we can test the model by comparing another sample to it. For instance, we can use the binomial distribution to predict the outcomes of flipping a fair coin 50 times, and compare sample results to the binomial model's predictions.

A different real-world activity might nearly meet the theoretical requirements of the Poisson distribution, but the relative frequencies of a single sample might not perfectly match the Poisson probabilities. In each case, we might want to ask whether observed data so closely follow

---

[1] Some authors prefer the term Chi-squared; this book follows the Minitab terminology.

the theoretical distribution that we could use the theoretical distribution as a model of the activity.

Tests that help to answer such questions are called *goodness of fit* tests. In earlier sessions, you have seen the Anderson-Darling Normality Test, which helps us to decide whether a sample has been drawn from a normal population. The tests in this lab are also goodness of fit tests, all of which rely on the *Chi-square* distribution.

## *A First Example: Simple Genetics*

In the 1860's, Gregor Mendel conducted a series of experiments, which formed the basis for the modern study of genetics. In several sets of experiments on peas, Mendel was interested in the heredity of one particular characteristic—the texture of the pea seed. He had observed that pea seeds are always smooth or wrinkled.

Mendel determined that smoothness is a *dominant* trait. In each generation, an individual pea plant receives genetic material from two parent plants. If either parent plant transmitted a smoothness gene, the resulting pea seeds would be smooth. Only if both parents transmitted wrinkled genes would the offspring pea be wrinkled.

This is the logical equivalent of flipping two coins, and asking about the chances of getting two heads. If the "parent" peas each have one smooth and one wrinkled gene (SW), then an offspring can have one of four possible combinations: SS, SW, WS, or WW. Since smoothness dominates, only the WW pea seed will have a wrinkled appearance. Thus, the probability of a wrinkled offspring is just 0.25.

Over a number of experiments, Mendel assembled data on 7324 second-generation hybrid pea plants. If the model just described is correct, we would expect one-fourth of the plants (1831) to be wrinkled, and the remaining 5493 to be smooth. In his trials, Mendel found 5474 smooth plants, and the rest were wrinkled.

We can use a Chi-square goodness of fit test to see whether the model accurately predicts the laboratory findings. Our null hypothesis is that the peas *do* follow Mendel's prediction:

$H_0$: $\pi_{wrinkled}$ = .25, $\pi_{smooth}$ = .75
$H_A$: $H_0$ is false

The Chi-square Goodness of Fit test compares the observed sample frequencies to the *expected frequencies* that we would find in a sample of the same size, if the hypothesized percentages are accurate. Hence, if the probability of *wrinkled* is 0.25, then we expect to find 1831 wrinkled pea plants in a sample of 7324 plants tested.

Above, we have the observed and the expected frequencies. We can use Minitab to help conduct the test.

🖱 In the blank worksheet, label the first three columns **Texture**, **Observed**, and **Expected**.

🖱 Enter the data, as shown in this completed worksheet below:

| | C1-T | C2 | C3 | C4 |
|---|---|---|---|---|
| | Texture | Observed | Expected | |
| 1 | Smooth | 5474 | 5493 | |
| 2 | Wrinkled | 1850 | 1831 | |
| 3 | | | | |

Worksheet 1 ***

In this test, the test statistic is given by this formula:

$$\chi^2 = \sum_{i=1}^{k} \frac{(o_i - e_i)^2}{e_i}$$

where

$o_i$ is the observed frequency of the $i$th category,

$e_i$ is the expected frequency of the $i$th category and

$k$ is the number of categories

The test statistic follows a chi-square distribution with $k - 1$ degrees of freedom. We can use Minitab's calculator feature to compute the test statistic, and the **Probability Distributions** function on the **Calc** menu to determine the significance level of the test statistic. We start by computing the test statistic, and storing it in the Minitab constant, **K1**.

---

💻 For goodness of fit tests, you may find it simpler to use Minitab to calculate the test statistic, and then consult a Chi-square table. In this example, we show how to do the entire test.

---

🖱 **Calc ➤ Calculator...** Type **K1** in the box marked **Store result in**. In the **Expression** box, copy the expression exactly as it is shown in the dialog on the next page.

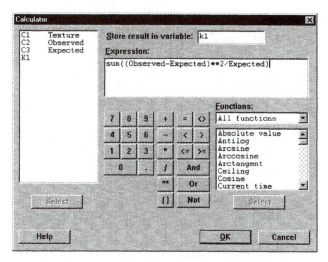

Nothing changes in the Session or Data Windows, but now the Minitab constant **K1** contains the value of the test statistic.

🖱 Open the Project Manager window (**Window ➤ Project Manager**) and click on the **Constants** folder icon to see the value of **K1** (0.26288).

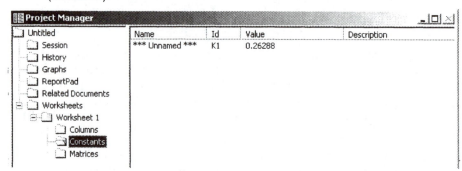

1. *If Mendel's model had perfectly predicted the frequencies of smooth and wrinkled seeds, what would the value of K1 be? Explain your reasoning.*

Now let's find the *P*-value for the statistic. The simplest approach is to find the cumulative probability between 0 and the test statistic; the *P*-value is the complement of this.

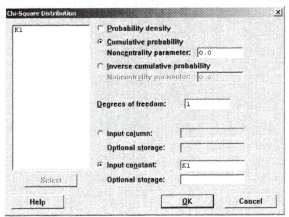 **Calc ➤ Probability Distributions ➤ Chi-Square...** Choose **Cumulative probability**. We need to specify the number of degrees of freedom, which is $k - 1$, or in this case, $2 - 1 = 1$. Enter **1** in **Degrees of freedom**. Choose **Input constant**, and type **K1**.

In the Session Window, we find the cumulative probability of the test statistic, which in this case is 0.3919. The *P*-value, therefore, is 0.6081, and we clearly fail to reject the null hypothesis. In other words, the experimental results are consistent with Mendel's model of inheritance, and present no persuasive challenge to his model.

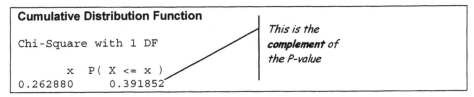

**Cumulative Distribution Function**

```
Chi-Square with 1 DF

       x   P( X <= x )
0.262880      0.391852
```

This is the **complement** of the P-value

## Testing for Independence

The Chi-square distribution is also useful for testing whether two qualitative or discrete variables are statistically independent of one another. Chi-square goodness of fit tests require multiple steps in Minitab. In contrast, tests for independence are quite straightforward.

The logic of the test is much the same as in the goodness of fit test. We start with a theory that predicts a set of frequencies. We then compare observed frequencies to the predicted ones. If the deviations between the observed and expected are sufficiently large, we reject the initial theory as incorrect.

For instance, in the **Student** dataset (open it now), we have one binary variable to distinguish males and females (**Gender**), and another which represents whether or not the student owns a dog (**Dog**). Given the archetypal image of "a boy and his dog" one might wonder if male undergraduates are more likely than their female classmates to be dog owners. We'll use a Chi-square test to decide if the data support or refute this theoretical connection between the variables. Our null hypothesis will be that dog ownership and gender are *not* related, and see if the data provide compelling evidence that they are.

Suppose that one-third of all students own dogs. If the null hypothesis is true (gender and dog ownership are unrelated), then one-third of the men own dogs, and one-third of the women own dogs. We understand that the sample results might not show precisely one-third for each group, and therefore anticipate some small departures from the theoretical expected values. Only if the deviations are sufficiently large will we reject the null.

For this test, the hypotheses are:

$H_0$: Dog ownership and gender are *independent*
$H_A$: $H_0$ is false

🖱 **Stat ➤ Tables ➤ Cross Tabulation and Chi-Square...** Select the two variables **Gender** and **Dog** (as shown below). Check **Counts** and then click the buttom marked **Chi-Square...**

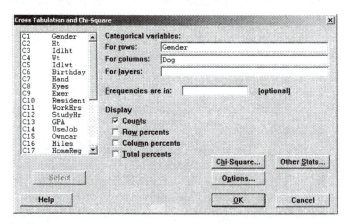

🖱 In the subdialog, check **Chi-Square analysis** and **Expected cell counts**, then **OK** to both dialogs.

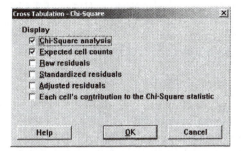

Look at the results in the Session Window. The rows of the table represent gender, and the columns represent dog ownership. Each cell of the table contains an *observed* joint frequency and an *expected* frequency. For example, there were 29 women who do not own dogs; under the null hypothesis, in a sample of 111 students, with 52 women in all and 64 non-owners, we would expect (52)(64)/111 = 29.98 female non-dog owners. Thus, there is a small discrepancy in that cell.

2.   *Look at the observed and expected frequencies in the four interior cells of the table (i.e. exclude the "All" cells). Do the discrepancies appear large or small to you? Explain your criteria for "large" and "small."*

```
Tabulated statistics: Gender, Dog
Rows: Gender   Columns: Dog

              No     Yes      All

Female        29      23       52
            29.98   22.02    52.00

Male          35      24       59
            34.02   24.98    59.00

All           64      47      111
            64.00   47.00   111.00


Cell Contents:        Count
                      Expected count

Pearson Chi-Square = 0.143, DF = 1, P-Value = 0.705
Likelihood Ratio Chi-Square = 0.143, DF = 1, P-Value = 0.705
```

The test statistic in this case has a value of 0.143, and under the null hypothesis would follow a Chi-square distribution with one degree of freedom. Compared to that distribution, the test statistic has a *P*-value of 0.705, which is far too large to reject the null hypothesis. Thus we conclude that there is not persuasive evidence of a relationship, and we persist in our null position that the two variables are *independent*.

## Sample Size Considerations

The chi-square test of independence comes with one *caveat*—it can be unreliable if the expected count in any cell is less than five. In such cases, a larger sample is advisable. For instance, let's look at another possible relationship in this worksheet. One question asked students to classify themselves as "below average," "average," or "above average" drivers. Let's ask if that variable is related to gender.

🖱 **Stat ➤ Tables ➤ Cross Tabulation and Chi-Square...** Select **Driver** and **Gender** as the variables.

Here is the output you will see:

```
Tabulated Statistics: Driver, Gender
Rows: Driver   Columns: Gender

                    Female    Male      All

Above Average           19      44       63
                     29.51   33.49    63.00

Average                 32      14       46
                     21.55   24.45    46.00

Below Average            1       1        2
                      0.94    1.06     2.00

All                     52      59      111
                     52.00   59.00   111.00


Cell Contents:       Count
                     Expected count

Pearson Chi-Square = 16.589, DF = 2
Likelihood Ratio Chi-Square = 16.992, DF = 2

* WARNING * 1 cells with expected counts less than 1
* WARNING * Chi-Square approximation probably invalid

* NOTE * 2 cells with expected counts less than 5
```

Take special note of the warning that appears at the end of the output. In this instance, the test results may be unreliable due to the uncertainty created by the cell with an expected count less than one, and the fact that two cells have expected counts less than five. In general, we should not draw inferences when expected counts are below five. As a simple description of this sample, though, it is appropriate to note that these particular men seem to have higher opinions of their own driving than the women do of theirs.

## Testing for Independence (Summary Data Only)

In the example just given, we started with case-wise data on a sample. Sometimes, we find a cross tabulation in another document, and might want to conduct the appropriate test. This is easily done.

For example, consider a local specialty mail-order company.[2] The firm routinely sends out thousands of catalogs to customers and to others who request them. Each time a new catalog is prepared, the marketing director runs one or more experiments in an effort to improve sales response. Consider the following scenario. One of the major costs in publishing catalogs is the cost of paper. The firm normally prints the catalog on a particular type of paper, but suppose a less costly paper might produce an equally appealing catalog. The marketing manager decides to print most of the catalogs on the conventional paper and also print a smaller run using a less expensive type of paper. He refers to the conventional paper as the "Champion" and the alternative paper as the "Challenger." As a way of gauging the appeal of the catalogs, tracks the number of orders generated by each mailing:

|                  | Challenger | Champion |
|------------------|-----------:|---------:|
| # mailed         | 24589      | 73721    |
| # orders placed  | 544        | 2053     |

As much as the marketing director would like to save money by starting to use the new less expensive paper, he will only do so if the "challenger" catalog does not diminish the sales response. In other words, he wants to avoid the error of cutting costs *and* reducing sales. Thus, unless the new paper proves itself a sales-reducer, he will make the change in paper type.

3. *Does it appear to you that the "Challenger" catalog is successful?*

---

[2] This example comes from a real firm that requested anonymity.

In Session 12, we learned a technique to compare two population proportions. Here, we'll see how to apply the Chi-square test of independence. The data for this example are in the worksheet **Catalog1**; open it now. As before, the hypotheses are these:

$H_o$: Paper and number of orders placed are *independent*
$H_A$: $H_o$ is false

4.   ***Look at the worksheet; how have we modified the summary information presented in the table above to create this worksheet?***

🖰   **Stat ➤ Tables ➤ Chi-Square Test (Table in Worksheet)...** Specify that the table is in columns C2–C3. The output (see below) is very similar to that shown earlier, except for some additional rows at the bottom.

---

**Chi-Square Test: Challenger, Champion**
Expected counts are printed below observed counts
Chi-Square contributions are printed below expected counts

|        | Challenger | Champion | Total |
|--------|-----------|----------|-------|
| 1      | 24045     | 71668    | 95713 |
|        | 23939.45  | 71773.55 |       |
|        | 0.465     | 0.155    |       |
|        |           |          |       |
| 2      | 544       | 2053     | 2597  |
|        | 649.55    | 1947.45  |       |
|        | 17.153    | 5.721    |       |
|        |           |          |       |
| Total  | 24589     | 73721    | 98310 |

**Chi-Sq = 23.494, DF = 1, P-Value = 0.000**

---

We see that the test statistic (23.494) has an associated *P*-value of approximately 0.000. Thus, we would reject the null hypothesis, and conclude that the number of orders placed *does* vary according to the paper used. In fact, closer inspection of the observed and expected frequencies indicates that the new paper under-performs the customary paper. As such, the Marketing director decided not to switch papers.

5.   ***If the Marketing director is making a statistical error in this case, what implications might the error have?***

## *Further Reading*

Details about Gregor Mendel's experiments and papers are available at
http://www.mendelweb.org/.

Agresti, Alan. *Categorical Data Analysis*. (NY: Wiley, 1990)

Kohler, Heinz. *Statistics for Business and Economics—Minitab Enhanced*.
(Cincinnati OH: Southwestern College, 2002), Chapter 14.

## *Moving On...*

Use the techniques of this lab session to respond to these
questions. *For each question, explain how you come to your statistical
conclusion, and suggest a real-world reason for the result.*

### Catalog2

This file is related to the example we just completed. In this case,
three experimental catalogs were shipped along with a control
(Champion) catalog. In the normal catalog, the company has a standard
schedule of shipping and handling charges, as well as a small fee to
insure the products in shipment. Customers normally have to judge the
various available product colors by relying on the colored inks using in
printing. The three experimental conditions were (1) reduced Shipping &
Handling charges, (2) reduced shipping insurance charges, and (3) an
offer of a free set of color samples with any order.

1. Is the number of orders placed independent of catalog
type?

2. What recommendations would you offer to the marketing
director before his next catalog mailing?

### Student

3. Is seatbelt usage independent of car ownership?

4. Is seatbelt usage independent of gender?

5. Is travel outside of United States independent of gender?

6. Is belt usage independent of smoking? Both variables may
gauge an individual's attitude towards risk; before
running the test, comment on what you expect to find.

## Mendel

This file contains summarized results for another one of Mendel's experiments. In this case, he was interested in four possible combinations of texture (smooth/wrinkled) and color (yellow/green). His theory would have predicted proportions of 9:3:3:1 (i.e., smooth yellow most common, one-third of that number smooth green and wrinkled yellow, and only one-ninth wrinkled green).[3] The first column of the dataset contains the four categories, and the second and fourth columns contain the respective observed and expected frequencies.

7. Perform a Goodness of Fit test to determine whether these data refute Mendel's null hypothesis of a 9:3:3:1 ratio.

8. Renowned statistician Ronald A. Fisher re-analyzed all of Mendel's data years later, and concluded that Mendel's gardening assistant may have altered the results to bring them into line with the theory, since *each one* of Mendel's many experiments yielded Chi-square tests similar to this and the one shown earlier in the session. Why would so many consistent results raise Fisher's suspicions?

## GSSGeneral

This file contains the data from the 1998 General Social Survey.

9. Does this sample of respondents provide statistically significant evidence that all signs of the Zodiac are not equally represented among U.S. adults?

10. Are the variables **Region** and **Race** independent?

11. Is there a statistical relationship between marital status and region?

12. Would you expect **Marital Status** and **Sex** to be independent or dependent? Why?

13. Run the test for independence for the variables **Marital Status** and **Sex** and report your findings. How might you explain the result?

---

[3] Kohler, Heinz (2002) *Statistics for Business and Economics—Minitab Enhanced.* (Cincinnati OH: Southwestern College) pp. 610–611.

## CPSVoting

This file contains data from the Current Population Survey concerning citizens' voting habits.

14. Are all geographic regions of the country equally represented in the Current Population Survey?

15. Did a person's employment status influence his or her likelihood of voting in the 1996 election?

16. Is there a significant relationship between a person's race and whether the person voted in the 1996 election?

## Violations

This worksheet contains records for 500 traffic violation citations issued at a small college.

17. For the semester in question, the student body followed this distribution: Freshmen–28%; Sophomores–25%; Juniors–26%; Seniors–21%. Do the traffic violation records reflect a significantly different pattern?

18. Are residency status and class independent?

19. Are class and fine amount independent?

20. Are class and type of violation independent?

# Analysis of Variance (I)

## Objectives

In this session[1], you will learn to:
- Perform and interpret a one-factor ANOVA
- Understand the assumptions necessary for a one-factor ANOVA procedures to yield reliable results
- Perform and interpret *post-hoc* tests for ANOVA
- Investigate the link between two-sample *t*-tests and ANOVA

## Comparing More than Two Means

In Session 12, we learned to perform tests that compare the means of two populations. In those tests our approach focused on the *difference between two means*. The null hypothesis was that the difference equaled zero (or some other value), and we examined the sample evidence about that difference.

Suppose we want to compare the means of *three or more* groups. Because we cannot meaningfully subtract three means from one another, we take an entirely new approach to the subject, thinking instead about *comparing variation within and among* different sample groups.

## A Simple Example

Consider a population of college students, some of whom work part-time to support themselves while in school. We might hypothesize

---

[1] I thank Professor Jane Gradwohl Nash for her significant contributions to the content and approach of this chapter.

that those who work many hours outside of school would find their grades suffering. To test that theory, we might use some sample data like that found in our **Student** file. Open that worksheet now.

One of the variables is called **Work**, and it is a categorical variable reflecting the number of hours per week that a student works part-time. The variable distinguishes among three groups of students: those who don't have a job, those who work fewer than 20 hours, and those who work 20 hours or more.

Our theory is that part-time work might adversely affect GPA; as in all hypothesis tests, the null hypothesis is that our theory is incorrect. Formally, we state the hypotheses as follows:

$$H_o: \mu_1 = \mu_2 = \mu_3$$
$$H_A: H_o \text{ is false} \quad \{\text{At least one mean is different}\}$$

Let's begin with an exploration of our sample data, computing group means and looking at the distributions of GPA for each of these three new groups in our sample. Do the following:

> 🖱 **Stat ▶ Basic Statistics ▶ Display Descriptive Statistics...** Select the variable **GPA**, and under **By variables**, select **Work**. Also, click on **Graphs** and request a **Boxplot of data**.

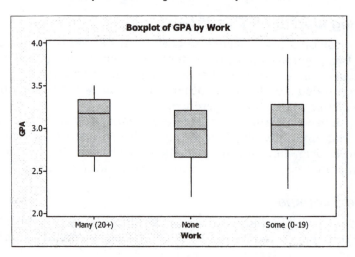

The boxplot shows some variation across the groups, with the highest GPA's belonging to students who work between 1 and 19 hours. As you look in the Session Window, you should notice that the mean

GPA differs slightly among the groups, but then again, the sample means of *any* three groups of students would differ to some degree.

Thus, we are left again with the essential inferential question. Should we ascribe the observed differences to sampling error, or do they reflect genuine differences among the three populations? Neither the boxplot nor the sample means offer conclusive evidence.

Part of the reason we perform a formal statistical test like ANOVA is to quantify the ambiguity. The ANOVA procedure will distinguish the extent of the variation we can ascribe to ordinary sampling error, and the variation we can ascribe to the *factor*, that is, hours of work.

As noted earlier, we initially hypothesize no difference among the three groups. Our alternative hypothesis states that at least one group has a different mean. As with other inferential procedures, a single-factor (or *one-way*) ANOVA provides reliable results under certain conditions. Although we can run the procedure with nearly any set of data, we can trust the results only if the following are true:[2]

- The samples are drawn independently
- The parent populations are normally distributed
- The populations share equal variances (*homoscedasticity,* or *homogeneity of variance*)

Before performing the test, we should consider whether these assumptions are valid for this set of data. The assumption of independent samples refers to the manner in which the samples were collected; here, we surveyed a large sample of students, and respondents indicated how much they work.

1. ***Under what circumstances might samples such as these not satisfy the independence assumption? Use your imagination to describe a scenario in which independence would be compromised.***

We can formally test the normality and equal-variance assumptions. With large samples, small departures from normality are acceptable, but equal variances assumption is more critical.

To test normality, we'll create a Graphical Summary for each group and consult the Anderson-Darling normality test.

---

[2] Note that ANOVA compares the centers (means) of $k$ different distributions. The comparison is reliable when we can assume that the distributions share the same shape and dispersion (spread). If any of the assumptions is violated, we should not use ANOVA, but instead use one of the techniques introduced in Session 21.

🖰 **Stat ➤ Basic Statistics ➤ Graphical Summary...** Create summaries for **GPA** by **Work**.

Recall that the Anderson-Darling statistic tests the null hypothesis that the sample was drawn from a normal population. We reject that null hypothesis only when the *P*-value is small.

2. *Look at the results for the three groups. Is there compelling evidence that any of the three is not normally distributed?*

Let us proceed to the next assumption, even if we have some reservations about normality for one sub-group. We have seen a test for the equality of two variances before. To compare more than two variances, we use a similar procedure.

🖰 **Stat ➤ ANOVA ➤ Test for Equal Variances...** Complete the dialog by selecting **GPA** as the **Response** and **Work** as the **Factor**. The results are shown here.

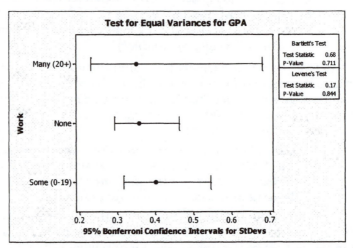

The graph summarizes the results of the test; you should look in your Session Window for additional information. For both Bartlett's and Levene's tests, the null hypothesis is that the three variances are equal. In this case, there is no compelling evidence that the variances are unequal, so we'll proceed to the ANOVA procedure.

To perform the ANOVA in this case, do the following:

🖰 **Stat ➤ ANOVA ➤ One-way...**  The **Response** variable is **GPA**; the **Factor** is **Work**.

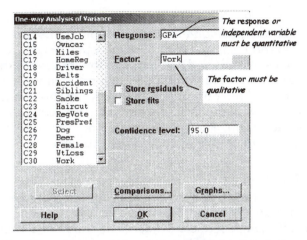

The ANOVA output is best understood as consisting of two parts. The upper portion is the typical ANOVA table, similar to that shown in your textbook. Here, the test statistic ($F$) equals 1.28, with a $P$-value of 0.282. We do not reject the null hypothesis because $P$ is so large. In other words, this is insufficient evidence to conclude that there is at least one meaningful difference in mean GPA among the three groups of students, despite the differences among the sample means.

## One-way ANOVA: GPA versus Work

```
Source    DF      SS      MS      F      P
Work       2   0.360   0.180   1.28   0.282
Error    106  14.928   0.141
Total    108  15.289

S = 0.3753   R-Sq = 2.36%   R-Sq(adj) = 0.51%

                                    Individual 95% CIs For Mean Based on
                                    Pooled StDev
Level          N    Mean    StDev   ---+---------+---------+---------+------
Many (20+)    12  3.0583   0.3513   (------------------*-----------------)
None          56  2.9504   0.3593   (-------*-------)
Some (0-19)   41  3.0673   0.4022            (---------*--------)
                                    ---+---------+---------+---------+------
                                    2.88      3.00      3.12      3.24

Pooled StDev = 0.3753
```

The lower portion of the ANOVA output shows the sample size, mean, and standard deviations for each of the three subgroups, as well as 95% confidence interval estimates for the three means. Note the extent to which they overlap, reflecting the degree of sampling error here.

Also note the "Pooled St. Dev." in the lower left. ANOVA is based on an assumption that the *variances* of the populations are equal; thus, we can regard the three sample variances as estimates of the common variance. The pooled standard deviation (the square root of the variance) combines the data from all three groups into one common estimate.

In the center of the output is a line showing three statistics: S, R-Sq, and R-Sq (adj). S is the overall standard error, equal here to the pooled standard deviation. The two R-square statistics are goodness of fit measures; we'll have more to say about these statistics in Session 16.

## *Another Example*

Analysis of variance is often the appropriate tool to analyze experimental data. In this example, we consider an experiment about the cardiovascular effects of exercise. Students at Ohio State University asked subjects to step up and down on a step in time with an electric metronome. Subjects were randomly assigned to one of three different stepping rates. Experimenters then counted the subjects' pulse rate.[3] The data are in the file called **Stepping**. Open it now.

For this example, we will treat **HR** (heart rate) as the response and **Frequency** (stepping rate) as the factor. The three stepping rates are coded as 0 for slow (14 steps/minute), 1 for medium (21 steps/minute) and 2 for high (28 steps/minute).

3. *Using the same methods illustrated earlier, assess the validity of the normality and equal variance assumptions. Explain your conclusions.*

Return to the One-way ANOVA dialog, and select **HR** as the response and **Frequency** as the factor. The results show an *F*-statistic of 6.00 with a *P*-value of 0.007. We should reject the null hypothesis and conclude that at least one mean is different from the others.

4. *Look at the confidence intervals in the ANOVA output. Which mean(s) do you think are different?*

Strictly speaking, the ANOVA results don't tell us which mean is different. We need a further test (known as a *post hoc* test) to determine where the differences occur. Minitab offers four different *post hoc* tests, each suited to a different inferential goal. As an introduction these

---

[3] The experiment was more complex than described here. We will revisit is in the next chapter; for a full account, visit the Data and Story Library on the Web at http://lib.stat.cmu.edu/DASL/Datafiles/Stepping.html.

comparisons, we'll illustrate only one method; interested readers should consult the Minitab help system for more details.

🖑 Return to the One-way ANOVA dialog, and click **Comparisons....** You will see the sub-dialog shown here. Check the box next to **Tukey's**, and use the default error rate of 5%.

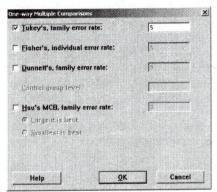

Tukey's method compares the means of each possible pair of factor levels (i.e., low vs. medium, low vs. high, medium vs. high). We obtain confidence intervals for the differences in means; intervals lying entirely above or below 0 indicate significant differences.

```
Tukey 95% Simultaneous Confidence Intervals
All Pairwise Comparisons among Levels of Frequency

Individual confidence level = 98.04%

Frequency = 0 subtracted from:

Frequency   Lower   Center   Upper   -----+---------+---------+---------+-
1          -11.17    8.40    27.97              (------*------)
2            7.13   26.70    46.27                  (------*------)
                                      -----+---------+---------+---------+-
                                         -25        0        25       50

Frequency = 1 subtracted from:

Frequency   Lower   Center   Upper   -----+---------+---------+---------+--
2           -1.27   18.30    37.87              (-------*------)
                                      -----+---------+---------+---------+--
                                         -25        0        25       50
```

The results in your Session Window are shown above, with one line highlighted. Our estimate of the difference between the level 0 and level 2 means is (7.13, 46.27). This is the only one of the three intervals indicating a meaningful difference. Thus, we would conclude that subjects in the slow-stepping treatment have lower heart rates than those in the fast-stepping treatment. If there are any other meaningful differences, this set of experimental data does not detect them.

## More on Normality

The one-way ANOVA command includes another way to evaluate the normality of the group distributions. We can do so by examining the *residuals* in our analysis. Residuals are the differences between the individual response values and the mean of the corresponding group. If each sub-group is normally distributed with the same variance, then the residuals will be normal as well. We can construct a normal probability plot of the residuals to evaluate their normality.

🖰  Return to the ANOVA dialog, and click **Graphs....** Under **Residual Plots**, select **Normal plot of residuals**.

5.  *Do the residuals indicate any violation of the normality assumption? Explain.*

## ANOVA and the Two-Sample t-Tests

Because we can use ANOVA to compare more than two means, you might wonder if it can also compare two means. In fact it can. Another factor in the stepping experiment was the height of the stair step. Half of the subjects used a 5.75-inch step and half used an 11.5-inch step. The heart rates of the two groups are approximately normal and have similar variances. The 2-sample *t*-test has these results:

```
Two-Sample T-Test and CI: HR, Height

Two-sample T for HR

Height   N    Mean   StDev   SE Mean
0        15   96.6   14.3    3.7
1        15   118.2  20.3    5.2

Difference = mu (0) - mu (1)
Estimate for difference:   -21.6000
95% CI for difference:   (-34.7253, -8.4747)
T-Test of difference=0 vs not =):T-Value = -3.37 P-Value = 0.002 DF = 28
Both use Pooled StDev = 17.5479
```

Now perform the equivalent ANOVA:

🖰  **Stat ➤ ANOVA ➤ One-way...** Select the variable **HR** as the response, and **Height** as the factor.

6. **Look at the results in the Session Window. At a 5% significance level, would you conclude that the two sub-populations share the same mean heart rate? What is the P-value?**

The P-value is identical to that of the two-tailed pooled-variance t-test. In a sense, this is not surprising, because these are equivalent tests, being performed with the same data. As such, they should give the same results. What is more, look closely at the two test statistics: $F = 11.36$ and $t = -3.37$. Though the rounded values disguise the fact, $t$ is the square root of $F$.

## Two Additional ANOVA Issues

You may have noticed the **One-way (Unstacked)** command in the ANOVA menu. If a worksheet is organized with the independent sample data in separate columns, we would use this command to perform the ANOVA. However, the command does not offer the *post hoc* tests, nor is there a simple way to conduct the tests to check equality of variances. Therefore, one is well advised to reorganize the worksheet to stack the data columns.

More substantively, one might also wonder what happens if we can identify more than one possible factor that influences the response variable. For example, in the stepping experiment, both the height of the step and the frequency of stepping may have contributed to differences in subjects' heart rates. The next session considers situations in which there are multiple qualitative factors.

## Moving On...

Use the techniques of this session to respond to the following questions. **Check the underlying ANOVA assumptions where possible using appropriate techniques; also, explain both the statistical evidence and theoretical reasons for your responses to the questions**. If any of the ANOVA assumptions is not satisfied, you should not proceed with the analysis. When you find significant differences, identify where the differences lie.

### JFKLAX

This file contains flight departure and arrival information for all flights between New York's Kennedy and Los Angeles International airports for a randomly selected day.

1. Do departure delays vary by airline (**Carrier**)?

2. Do arrival delays vary by airline?

## Ohio Votes

This worksheet holds the results of the 2000 Presidential election for the state of Ohio. For the sake of this example, let's consider Ohio to be a randomly-selected state.

3. Was the voter turnout percentage related to which candidate won the county? Speculate about the possible reasons for the result that you have found.

## Mft

This dataset contains Major Field Test (MFT) results, SAT scores and GPA's for a group of college seniors majoring in a science. Department faculty members are interested in predicting a senior's MFT performance based on their high school or college performance. The variables **GPAQ**, **VerbQ**, and **MathQ** indicate the quartile in which a student's GPA, Verbal SAT, and Math SAT scores fall within the sample.

4. Do mean total scores on the MFT vary by GPA quartile? Comment on distinctive features of this ANOVA.

5. Does the relationship between total score and GPA hold true for each individual portion of the MFT?

6. Do mean total scores on the MFT vary by Verbal SAT quartile?  Math SAT quartile?

7. Based on these results, one faculty member suggests that College grading policies need revision. Why might one think that, and what do you think of the suggestion?

## Milgram

These are data from several famous obedience studies conducted by Dr. Stanley Milgram.[4] His studies, inspired by questions arising from the Holocaust, became extremely controversial. Milgram wanted to understand why people so often obey authority, even when asked to do things they would not do on their own.

---

[4] Milgram, Stanley. *Obedience to Authority* (New York: Harper, 1975).

Each experiment involved three people: Dr. Milgram, a male subject, and another male colleague of Dr. Milgram. Subjects were told that they would be teaching another person (the accomplice). Specifically, the teacher would ask the learner questions while the learner was connected to a device that the subject believed was capable of delivering electric shocks of varying intensity. Dr. Milgram instructed the teacher to shock the learner for each incorrect answer, and to increase the intensity of the shock for each successive error. The teacher controlled a dial marked in increments from 15 volts (very mild) to 450 volts (severe). At the outset of the experiment, Milgram administered a 45-volt shock to the teacher. This was the only shock that anyone received—the "shock generator" did not actually do anything at all to the learner. The point *of the experiment was to observe the behavior of the teacher while he believed* he was shocking his student.[5]

The response variable is the maximum number of volts that a subject "administered" to his student. Milgram ran four different variations of the experiment, modifying the physical proximity of the teacher and learner. These variations are identified in the dataset as Experiments 1, 2, 3, and 4.

8. Although these experiments were run as separate studies, we could consider them to be four independent samples of a single response variable with four treatments. Do the data meet the necessary assumptions to conduct an ANOVA? Explain your conclusions.

## GSSEduc

This worksheet contains data from the 1998 General Social Survey. All variables related to subjects' education.

9. Does the number of years of schooling vary by race?

10. Does the number of years of schooling vary according to the level of education attained by a subject's father?

## Haircut

These data are extracted from the Student dataset. Students reported the price they paid for the most recent professional haircut.

---

[5] These experimental conditions caused the subjects so much distress that they would no longer be permitted today under the ethical guidelines established by the American Psychological Association.

11. Do haircut prices vary by region?

12. Do men and women pay comparable prices for haircuts?

## GSSGeneral

This worksheet has more data from the 1998 General Social Survey.

13. We have data in this worksheet about respondents' ages and the region in which they live. Suggest some reasons that we might expect the mean age of adults to vary across different parts of the United States.

14. Is the mean age of adults the same in each region of the United States?

## Hotdog

This worksheet contains calorie and sodium content data for a sample of three types of hotdogs.

15. Are there statistically significant differences in the caloric content among the three types of hotdogs?

16. Are there statistically significant differences in the sodium content among the three types of hotdogs?

# Analysis of Variance (II)

## *Objectives*

In this session[1], you will learn to:
- Perform and interpret two-way ANOVA
- Analyze designs with more than two factors
- Understand and interpret main effects
- Understand and interpret interaction effects

## *Going Beyond a Single Factor*

In Session 14 we learned to draw inferences about possible relationships between a quantitative response and a qualitative factor. In both observational and experimental studies we sometimes want to investigate the influence of several factors. What is more, we may want to understand the simultaneous impact of those factors. In other words it is not sufficient to test each factor one at a time; we need to test their individual and collective effects all at once.

## *A Two-Way ANOVA*

In a two-way ANOVA, we hypothesize that there are two categorical factors that might influence the value of the response variable. In the previous session we considered one aspect of an experiment conducted by students at Ohio State University. In fact, the experiment was more complex than presented in Session 14. Let's return to that experiment now.

---

[1] I thank Professor Jane Gradwohl Nash for her significant contributions to the content and approach of this chapter.

In the experiment, the students wanted to understand the effects of physical activity on subjects' heart rates. They asked subjects to step up and down on a step at a particular rate of speed, controlled by an electric metronome. Thirty subjects were randomly assigned to one of three different stepping rates. In addition, the experimenters had two steps of different heights. Thus, we have a quantitative response variable (heart rate) and *two* qualitative factors: Frequency (stepping rate) and Height (stair height). The experimenters used three stepping rates and two different step heights.

This design created six distinct experimental conditions, and five subjects were assigned at random to each condition. We might visualize the experimental design as a table in which each cell represents one of the experimental conditions:

| | Frequency | | |
|---|---|---|---|
| **Height** | Low | Med | High |
| Low | 5 | 5 | 5 |
| High | 5 | 5 | 5 |

This design will allow us to investigate two kinds of relationships. First, we can ask about *main effects*, or the separate impact of step height and stepping frequency on subjects' pulse rates. Second, we can ask about *interaction effects* or the combined impact of the two factors on pulse rates. For instance we might find that step height does not matter for low-frequency steppers, but becomes influential for fast-steppers.

In general terms we have factor A which might have *a* distinct levels (values), and factor B with *b* levels. We can hypothesize that the response variable has equal means for all levels of A, and equal means for all levels of B. In addition we can hypothesize that the response variable has different means for the *interaction* of A and B. Unlike the single-factor ANOVA in a two-way analysis *we must have equal numbers of observations for each of the a x b treatments* or cells. This is an important difference between one- and two-way ANOVA.

We can use the two-way ANOVA (also known as the two-factor ANOVA) command to analyze this data set. Please note that this Minitab command requires that we have equal numbers of observations in each cell; this condition is quite manageable in a designed experiment, but more restrictive in an observational study. We'll see later how to proceed with if we have unequal numbers of observations.

Open the **Stepping** worksheet, and do the following:

🖱 **Stat ➤ ANOVA ➤ Two-way...** Select **HR** as the **Response**, with **Frequency** as the **Row factor** and **Height** as the **Column factor**. Check **Display means** for both factors.

🖱 Under **Graphs** select **Normal plot of residuals**. Compare your entries to the dialog box shown on the facing page. You may want to review the short discussion of residuals in Session 14.

1. *If the residuals are normally distributed, we can safely assume that the response variable is normal for both factors. Do you think we can assume normality? Explain.*

Behind the normal plot of residuals, you should see the following in the Session Window.

```
Two-way ANOVA: HR versus Frequency, Height

Source        DF       SS       MS      F      P
Frequency      2    3727.8  1863.90   9.55  0.001
Height         1    3499.2  3499.20  17.93  0.000
Interaction    2     210.6   105.30   0.54  0.590
Error         24    4683.6   195.15
Total         29   12121.2

S = 13.97   R-Sq = 61.36%   R-Sq(adj) = 53.31%

                    Individual 95% CIs For Mean Based on
                    Pooled StDev
Frequency   Mean    --------+---------+---------+---------+-
0           95.7    (-------*------)
1          104.1           (-------*------)
2          122.4                           (-------*-------)
                    --------+---------+---------+---------+-
                          96        108       120       132

                    Individual 95% CIs For Mean Based on
                    Pooled StDev
Height   Mean       -+---------+---------+---------+--------
0        96.6       (-------*------)
1       118.2                        (------*-------)
                    -+---------+---------+---------+--------
                    90        100       110       120
```

In the ANOVA table we find three $F$ ratios, corresponding to the two factors and to their interaction.

2. *Which if any of the three effects listed is statistically significant? Explain.*

The ANOVA output indicates that there is no significant interaction between the two factors. We infer this from the *P*-value of 0.590. In other words, each of the two factors makes a unique and independent contribution to a subject's heart rate. We can now proceed to consider those main effects. In our next example, we will see a significant interaction, and learn to interpret its meaning.

Notice the means and confidence intervals displayed below the ANOVA table. We interpret them just as we did in the previous session.

3. ***Which seems to have a greater impact on heart rate: step height or step frequency? Explain.***

We can display the magnitude of the main effects graphically to help visualize the varying impacts on heart rate, as follows.

🖱 **Stat ➤ ANOVA ➤ Main Effects Plot...** Again, select **HR** as the response and **Height** and **Frequency** as the factors.

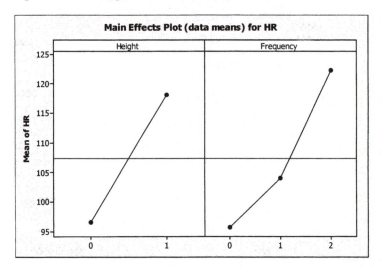

The left panel shows the mean heart rates for the low- and high-stepping subjects. The right panel shows three heart-rate means, corresponding to the three different stepping frequency groups. It appears that the frequency means vary more than do the height means.

## A More Complex Example

Session 1 introduced a classroom experiment using paper helicopters. The goal of that experiment was to identify those design features that influence the flight duration of a helicopter.[2] In the first Session, we analyzed a small subset of the data. Now we are ready to take a more sophisticated look at the full data set.

In the experiment, we considered three factors simultaneously:

- Wingspan: the length of the wings. The levels tested were 3 inches and 4 inches.
- Body width: the width of the folded lower portion of the helicopter. The levels tested were 1 inch and 2/3 inch.
- Clip: presence or absence of a paperclip at the base of the helicopter.

There are eight different ways to combine these three factors, and the students flew each unique design eight times, recording the flight time for each flight. In all, the class ran 64 different test flights as indicated. The worksheet **Helicopters** contains the experimental design and the flight durations for all 64 runs. Open that worksheet now.

Although there are three factors in this experiment, let's begin with a two-way ANOVA focusing on the second and third factors. We'll add the third factor to our analysis shortly.

🖱 **Stat ➤ ANOVA ➤ Two-way...** The response variable in this worksheet is called **Duration**. Choose **BodyWidth** as the **Row factor** and **Clip** as the **Column factor**.

Look closely at the output, a portion of which is shown here:

**Two-way ANOVA: Duration versus BodyWidth, Clip**

| Source | DF | SS | MS | F | P |
|---|---|---|---|---|---|
| BodyWidth | 1 | 2.2877 | 2.28766 | 2.89 | 0.094 |
| Clip | 1 | 2.8308 | 2.83081 | 3.57 | 0.064 |
| Interaction | 1 | 5.6525 | 5.65251 | 7.14 | 0.010 |
| Error | 60 | 47.5174 | 0.79196 | | |
| Total | 63 | 58.2883 | | | |

The main effects are significant at the .10 level, but not at the .05 level. However, the interaction has a very low $P$-value of 0.01. What does this mean? It means that the impact of body width on flight duration *depends* on whether a helicopter has a paperclip. In other words, the main effects of one factor vary in the presence of the other factor. For this reason, it is very important that we consider and interpret significant interactions before we analyze the main effects. As always, a graph can help deepen our understanding.

---

[2] See Session 22 for a detailed discussion of the design of this experiment.

🖱 **Stat ➤ ANOVA ➤Interactions Plot...** Complete the dialog as shown, to create the graph displayed below.

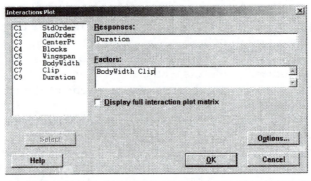

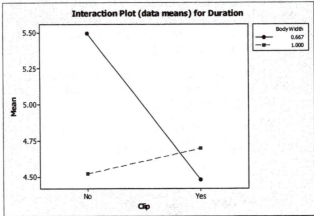

This plot displays all three variables. The vertical axis represents the mean flight durations. Black dots correspond to the 32 helicopters with narrow bodies (2/3 of an inch), and the red squares correspond to the other 32 helicopters with wide bodies (1-inch). The horizontal axis represents the binary paperclip variable.

In the absence of a paperclip, we can see that narrow-bodied helicopters fly longer than wide-bodied models. However when there is a paperclip on the helicopter, the difference evaporates, and the wide-bodied helicopters actually fly a bit longer.

## *Beyond Two Factors*

Because this experiment was formally designed using three factors, we really should analyze the data in a way that accounts for all three factors at once. This will take us a bit beyond the usual limits of two-way ANOVA, but ultimately this will help to deepen your understanding of the technique. Because this worksheet was initially created using a Minitab Design of Experiments (DOE) command[3] we can do the following:

🖑 **Stat ➤ DOE ➤ Factorial ➤ Factorial Plots...** Check all three plots, as shown. Click the **Setup** button for each. Specify that **Duration** is the Response and **Wingspan**, **BodyWidth**, and **Clip** are the factors.

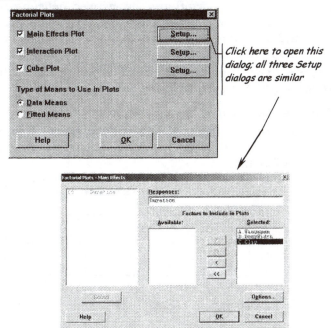

On the next page we see the Main Effects Plot. This illustrates the difference in mean flight duration for each of the factors considered separately.

---

[3] See Session 22 for full details.

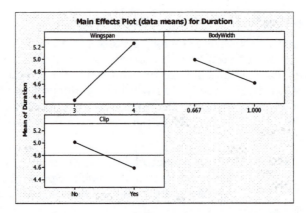

The upper left panel shows that short-winged helicopters had mean flights of about 4.4 seconds, and long-winged ones had mean flights exceeding 5.2 seconds. If this were all the information available to us, we would conclude that longer wings lead to longer flights.

**4. Interpret the upper right panel. What is the main effect of body width on flight duration?**

**5. What is the main effect of the paperclips?**

Next we see a plot of interactions between pairs of factors. Here again we have three different graphs, each one showing the mean flight duration for a *pair* of factors.

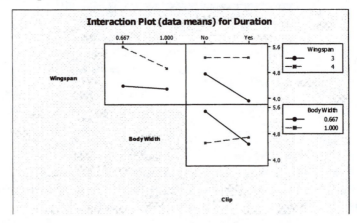

The top row shows the interaction between wingspan and the other two factors. The second row illustrates the interaction that we saw earlier between body width and the presence of a paper clip. Thus, the left graph in the top row displays the mean flight times accounting simultaneously for wingspan and body width. The dashed line represents long wings, and the solid line short wings. We see four means, corresponding to the four endpoints of the two lines. We see a possible mild interaction between the two factors: for long-winged helicopters, body width seems to influence flight duration more dramatically than it does for short-winged helicopters.

**6.   *What does the right-hand graph in the top row tell you?***

We should begin our interpretations of experimental results by first looking for interactions. Why? When we looked at the main effects plots, we saw substantial main effects associated both with wing length and body width. We now know, however, that the magnitude of the effect of these variables depends on the state of other factors. Thus, we need to be very careful in drawing conclusions about main effects until we have investigated the possibility of interactions.

The Cube Plot summarizes the flight duration data taking all three factors into account at once. The four corners of the "ceiling" of the cube correspond to the average flight times for wide-body helicopters (**BodyWidth** = 1 inch) and the corners of the "floor" correspond to the flight times for the four narrow-body models (**BodyWidth** = .667 inch). The left and right walls of the cube represent the conditions of short and long wings; the front and back faces of the cube represent the absence and presence of a paperclip.

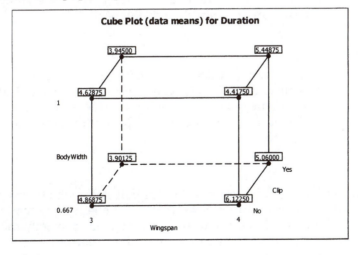

7. *Find the value 4.62875 seconds in the upper left corner of the front face of the cube. Which helicopter model does it represent?*

8. *Which helicopter model had the longest mean flight duration in this experiment?*

## Moving On...

Use the techniques of this session to respond to the following questions. As in the prior session, check the underlying ANOVA assumptions where possible using appropriate techniques. Also, explain both the statistical evidence and theoretical reasons for your responses to the questions. If any of the ANOVA assumptions is not satisfied, comment on the implications for further analysis.

### Falcon

This worksheet contains the results of a study to investigate the residual effects of DDT among falcons. DDT is a pesticide that was banned in the United States due to its long-lasting detrimental effects on bird populations. Years later, residues of DDT were still present in birds.

1. Using a two-way ANOVA, does the evidence suggest a relationship between the amount of DDT residue and the age of the bird? The nesting site? Is there a significant interaction between the two factors?

### Helicopters

2. Use the two-way ANOVA commands to analyze the effects of wingspan and body width on flight duration. Fully discuss the degree to which the ANOVA assumptions are satisfied, and report and interpret both main effects and interactions.

### Haircut

This worksheet comes from the Student data, which was collected on the first day of class. Students were asked the last price they paid for a professional haircut. In addition, they were asked to specify the region (rural, suburban, or urban) where they had their hair cut.

3. Is the price of a haircut related to a person's sex, the region where they paid for the haircut, or some

combination of these two factors? Explain your conclusions referring to appropriate statistical analysis.

## Student

Recall that these data are collected from the first day of a business statistics class. Because we could not assign individual students to particular factor levels, we cannot assume equal numbers of students in each cell for a two-way ANOVA. Thus, the two-way ANOVA command will not operate. However, we can construct main effects and interaction plots for the following questions.

4. Propose a theory to explain why both gender and whether a student smokes might affect the student's GPA. Using this worksheet test your theory and discuss your findings.

5. Do gender and one's rating of personal driving ability affect the number of accidents that a student reported in the survey? Comment on noteworthy features of this analysis.

## GSSEduc

This worksheet contains General Social Survey data pertaining to the educational attainment of respondents. Because we cannot assign individual respondents to particular factor levels, we cannot assume equal numbers of respondents in each cell for a two-way ANOVA. Thus, the two-way ANOVA command will not operate. However, we can construct main effects and interaction plots for the following questions.

6. Investigate the extent to which the number of years of schooling (**EDUC**) depends on the sex and race of the respondent.

7. Investigate the extent to which the number of years of schooling (**EDUC**) depends on the sex and highest degree earned by the respondent's father (**PADEG**).

## Swimmer1

These data reflect practice times for a high school swim team. There was no formally designed experiment here, so we cannot assume equal numbers of observations in each cell for a two-way ANOVA. We can construct main effects and interaction plots for the following questions.

8.  Investigate the extent to which swimmers' times depend on their gender and whether the time was for the first or second heat.

9.  Investigate the extent to which swimmers' times depend on the particular event and their gender.

# Session 16

## Linear Regression (I)

### *Objectives*

In this session, you will learn to:

- Perform a simple two variable linear regression analysis
- Test hypotheses about the relationship between two quantitative variables
- Evaluate the Goodness of Fit of a linear regression model

### *Linear Relationships*

Some of the most interesting questions of statistical analysis deal with the relationships between two quantitative variables: How much will sales increase if we spend more on advertising? How much will regional water consumption increase if the population increases by 1,000 people? How much more heating fuel will I use if I add a room to my house?

In each of these examples, there are two common elements: a pair of quantitative variables (e.g., sales revenue and advertising expense), and a *theoretical reason* to expect that the two variables are somehow related.

Linear regression analysis is a tool with several important applications. First, it is a way of *testing hypotheses* concerning the relationship between two numerical variables. Second, it is a way of *estimating* the specific nature of such a relationship—beyond asking "are sales and advertising related?" regression allows us to ask *how* they are related. Third, and by extension, it allows us to *predict* values of one variable if we know or can estimate the other variable.

As a first illustration, consider the classic economic relationship between consumption and income. Each additional dollar of income enables a person to spend (consume) more. As such, when income increases, we expect consumption to rise as well. Similarly a person with more income than I have will, other things being equal, tend to consume more than I do. Let's begin by looking at aggregate income and consumption of all individuals in the United States over a long period of time.

🖰 **File ➤ Open Worksheet...** Select the file **US**. This file contains different economic and demographic variables for the years 1965–2000. We are interested in **PersCon** and **PersInc**, which represent aggregate personal consumption and aggregate personal income, respectively.

First, let's construct a scatterplot of the two variables. Our theory says that consumption depends on income. In the language of regression analysis consumption is the *dependent* or *response* variable. Income is the *independent* or *predictor* variable. It is customary to plot the dependent variable on the Y-axis, and the independent on the X-axis.

🖰 **Graph ➤ Scatterplot...** Select a simple scatterplot, and specify that the Y variable is **PersCon** and the X variable is **PersInc**. Click **OK**.

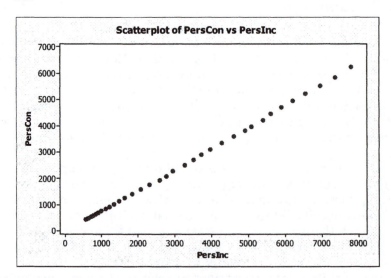

As you look at the resulting plot, you can see that the points fall into nearly a perfect straight line. This is an example of pronounced *positive* or *direct* relationship, and a good illustration of what a linear relationship looks like. It is called a positive relationship because the line has a positive, or upward, slope. One interpretation of the phrase "linear relationship" is simply that X and Y form a line when graphed. But what does that mean in real-world terms? It means that Y changes by a constant amount every time X increases by one unit.

In this graph, the pattern formed by the points is nearly a perfect line. The regression procedure will estimate the equation of that line which comes closest to describing the pattern formed by the points.

🖰 **Stat ➤ Regression ➤ Regression...** The **Response** variable is **PersCon**, and the **Predictor** is **PersInc**.

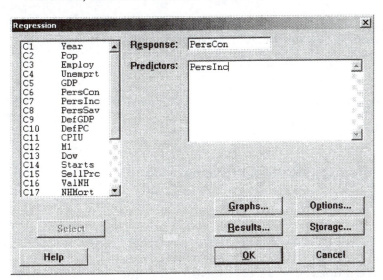

Since the output from the regression procedure is fairly involved, the output from this example is shown below. On your screen, maximize the Session Window, scroll back to the words "Regression Analysis," and follow this discussion while checking the output on your screen.

The regression output consists of five parts: The estimated equation, the table of coefficients, the goodness of fit measures, the ANOVA table, and a table of unusual observations. Your text may deal with some or all of these parts in detail; in this session, we'll take them up one at a time.

The uppermost part is the estimated equation. Minitab estimates that the line that most nearly fits these points is given by the equation

$$\text{PersCon} = -48.0 + 0.800 \text{ PersInc}$$

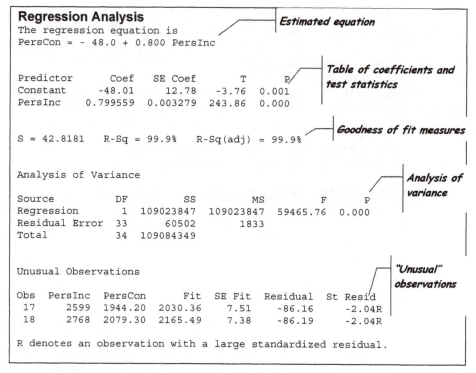

**Regression Analysis**
```
The regression equation is                            Estimated equation
PersCon = - 48.0 + 0.800 PersInc


Predictor        Coef    SE Coef       T       P      Table of coefficients and
Constant       -48.01      12.78   -3.76   0.001      test statistics
PersInc      0.799559   0.003279  243.86   0.000


S = 42.8181    R-Sq = 99.9%   R-Sq(adj) = 99.9%       Goodness of fit measures


Analysis of Variance
                                                             Analysis of
Source            DF         SS         MS        F      P   variance
Regression         1  109023847  109023847  59465.76  0.000
Residual Error    33      60502       1833
Total             34  109084349


Unusual Observations                                        "Unusual"
                                                            observations
Obs  PersInc  PersCon      Fit  SE Fit  Residual  St Resid
 17     2599  1944.20  2030.36    7.51    -86.16     -2.04R
 18     2768  2079.30  2165.49    7.38    -86.19     -2.04R

R denotes an observation with a large standardized residual.
```

The slope of the line (.80) indicates that if Personal Income increases by 1 billion dollars, Personal Consumption increases by 0.8 billion dollars. In other words, in the aggregate, individuals consumed about 80 cents of each additional dollar earned.[1]

What does the intercept mean? The value of −48.0 literally says that if income were 0, consumption would be −48.0 billion dollars. This makes little sense, but reflects the fact our data set lies very far from the Y axis. The estimated Y-intercept is a huge extrapolation beyond the observed data. The line which we estimate must cross the axis somewhere; in this instance, it crosses at −48.0.

Now look at the table of coefficients. Notice that next to the word **Constant** is −48.01, and next to **PersInc** is 0.799559. These are the same values as in the equation above, except that they are carried to several

---

[1] You may recognize this as the *marginal propensity to consume*. Note that the slope refers to the *marginal* change in Y, given a one unit change in X. It is *not* true that we spent 80 cents of *every* dollar. We spend 80 cents of the next dollar we have.

more decimal places. Because this relationship is so unusually linear, another example will serve better to explain the rest of the output.

## A non-linear example

Besides consuming our income, we can also save or invest it. Let's examine the relationship between Income and Saving, and compare it to the regression we've just done.

🖰   **Graph ➤ Scatterplot...**  Create another simple plot, but this time select **PersSav** as Y, and **PersInc** as X.

1.   *How does this graph compare to the first scatterplot?*

2.   *Does there appear to be any kind of relationship?*

3.   *In general, are low X values associated with low or high Y values? Are high X values associated with low or high Y values?*

Clearly, the connection between Savings and Income is not nearly as strong (or as linear) as the relationship between Consumption and Income. We'll return to this example when we investigate non-linear relationships in Session 19. For now let's look at another linear relationship.

## Another linear relationship

Consider the relationship between the population of the United States and the number of people who are employed. In our dataset, these variables are measured as **Pop** and **Employ**.

4.   *In a sentence or two, explain (a) why these two variables might be related, and (b) what that relationship might look like.*

Create a scatterplot with population on the horizontal axis and employment on the vertical axis.

5.   *Describe what you see in the scatterplot. Does it confirm what you suspected about these two variables?*

6.   *How does this scatterplot compare to the first one we made?*

The fact that these points do not align perfectly means that this relationship is a bit weaker than the relationship between income and consumption. To measure the strength of linear relationships, we return to the correlation coefficient.

🖱 **Stat ➤ Basic Statistics ➤ Correlation...**  Select the variables **Pop** and **Employ**.

Though we did not compute it in our first example, the correlation between consumption and income is approximately 1.0.

> 7.   *What does it mean to say that population and employment have a correlation of less than 1?*

🖱 **Stat ➤ Regression ➤ Regression...** Now, the **Response** is **Employ**, and the **Predictor** is **Pop**.

> 8.   *Interpret the resulting equation. What does the slope of the line tell you about employment and population?*

## Inference from Output

If we were to hypothesize a relationship between population and employment, it would be positive: the more people living in the U.S., the more people employed in the economy. Formally, the theoretical model of the relationship might look like:

$$\text{Employment} = \text{Intercept} + (\text{slope})(\text{Population}) + \varepsilon \quad \text{or}$$
$$Y_i = \beta_0 + \beta_1 X_i + \varepsilon_i$$

If $X$ and $Y$ genuinely have a positive relationship, $\beta_1$ is a positive number. If they have a negative relationship, $\beta_1$ is a negative number. If they have *no relationship at all*, $\beta_1$ is zero.

When we estimate the line using the regression procedure, we compute an estimated slope. Typically, this slope is non-zero. It is crucial to recognize that the estimated slope is a result of the particular sample at hand. A different sample would yield a different slope. Thus our estimated slope is subject to sampling error, and therefore is a matter for hypothesis testing.

One of the standard tests we perform in a regression analysis is designed to judge whether there is any *significant linear relationship* between $X$ and Y. Our null hypothesis is that there is not; i.e., that the 'true' slope is zero:

$$H_0: \beta_1 = 0$$
$$H_A: \beta_1 \neq 0$$

```
The regression equation is
Employ = - 85167 + 0.804 Pop

Predictor      Coef  SE Coef       T      P
Constant     -85167     2968  -28.70  0.000
Pop         0.80373  0.01271   63.25  0.000
```

Now look at the table of coefficients for this regression. The right-most two columns are labeled T and p. These represent *t*-tests asking if the intercept and slope (respectively) are equal to zero.[2] In this case, the value of the test statistic for the slope is 63.25, and the *P*-value associated with that test statistic is approximately 0. As in all *t*-tests, we take this to mean that we should *reject* our null hypothesis, meaning *there is a statistically significant relationship* between X and Y.

We know that the regression procedure, via the *least squares method* of estimation, gives us the line which fits the points better than any other. We might ask just how "good" that fit is. It may well be the case that the "best fitting" line is not especially close to the points at all!

Another standard part of regression output is a statistic that addresses just that question. The statistic is called the *coefficient of determination,* represented by the symbol $r^2$. It is the square of $r$, the coefficient of correlation. Among the goodness of fit measures in the regression output, locate **R-sq**.[3] $r^2$ can range from 0 to 100%, and indicates the extent to which the line fits the points; 100% is a perfect fit, such that each point is on the line. Another interpretation of the coefficient of determination is that it is the percent of variation in Y explained by the variation in X. The higher the value of $r^2$, the better. In this case the fit is not perfect but is very nearly so.

```
S = 1748.02   R-Sq = 99.2%   R-Sq(adj) = 99.2%
```

To better visualize how this line fits the points, do the following:

🖰 **Stat ➤ Regression ➤ Fitted Line Plot...** Select **Employ** as the **Response, Pop** as the **Predictor**. Maximize the resulting graph to get a good look at the graph. According to these results

---

[2] In this example, the intercept has little practical meaning. Therefore, in this session, we bypass it. A later example discusses a hypothesis test concerning the intercept.

[3] You will also find R-sq (adj) next to R-sq. The *adjusted* $r^2$ is used in Multiple Regression, and we will discuss it in a later session.

about 99% of the variation in employment is associated with changes in the size of the population.

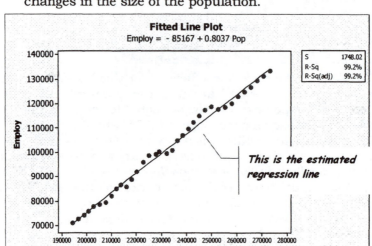

**Fitted Line Plot**
Employ = − 85167 + 0.8037 Pop

| S | 1748.02 |
| R-Sq | 99.2% |
| R-Sq(adj) | 99.2% |

*This is the estimated regression line*

## An Example of a Questionable Relationship

We've just seen two illustrations of fairly strong, statistically significant linear relationships, backed up by strong theoretical reasoning. Let's look at another example, where the theory is not as compelling.

Nutrition and session experts sometimes concern themselves with a person's body fat, or more specifically the percentage of total body weight made up of fat. Measuring body fat precisely is more complicated than measuring other body characteristics, so it would be nice to have a mathematical model relating body fat to some easily measured human attribute. Our dataset called **Bodyfat** contains precise body fat and other measurements of a random sample of adult men. Open the worksheet.

Suppose we wondered if height could be used to predict the percentage of body fat. We could use these data to investigate that relationship.

🖰 **Graph ➤ Scatterplot...** Create another simple scatterplot. Y is **FatPerc** and X is **Height**.

9.   *Is there evidence in this graph of a positive linear relationship?*

⤴ **Stat ➤ Regression ➤ Regression... FatPerc** is the response, and **Height** is the predictor.

10. *Do the regression results suggest a relationship? Which specific parts of the output did you use to decide?*

11. *What "real-world" reasons account for the observed results?*

12. *What does the intercept tell you in this regression?*

The key point here is this: Even though we can estimate a least squares line for any pair of variables, we may often find that there is no statistical evidence of a relationship. Neither the scatterplot nor the estimated slope is sufficient to determine significance; we must consult a *t*- or *F*-ratio for a definitive conclusion.

## An Estimation Application

In my home, we use natural gas for heating our house and our water, as well as for cooking. Each month, we receive a bill from the gas company for the gas we burn. The bill is a treasure trove of information, including two variables that are the focus of this example.

The first is a figure that is approximately equal to the number of cubic feet of natural gas consumed per day during the billing period. More precisely, it equals the number of "therms" per day; a therm is a measure of gas consumption reflecting the fact that the heating capacity of natural gas fluctuates during the year.

The second variable is simply the mean temperature for the period. These two variables are contained in the data file called **Utility**.

⤴ **File ➤ Open Worksheet...** Open the worksheet called **Utility**. The variables we want are called **GaspDay** and **MeanTemp**.

We start by thinking about *why* gas consumption and outdoor temperature should be related. Then we can think about *how* they might be related.

13. *Why might these two variables be related?*

14. *What would a graph of that relationship look like? Before proceeding, sketch the graph you expect to see.*

⤴ **Graph ➤ Scatterplot...** Construct a simple scatterplot with **GaspDay** on the vertical axis, and **MeanTemp** on the horizontal.

**15.** *Does there appear to be a relationship? How does it compare to your sketch?*

**Stat ➤ Regression ➤ Regression**... This time, you decide which variable is the response and which is the predictor.

Now look at the regression results.

**16.** *What do the slope and intercept tell you about the estimated relationship?*

**17.** *What does the negative slope indicate?*

**18.** *Is the estimated relationship statistically significant?*

**19.** *How would you rate the Goodness of Fit?*

One fairly obvious use for a model such as this one is to predict or estimate how much gas we'll use in a given month. For instance, in a month averaging temperatures of 40 degrees, the daily usage could be computed as

$$\text{GaspDay} = 15.5 - 0.215\,(40)$$
$$= 15.5 - 8.6$$
$$= \ \ 6.9 \text{ therms per day.}$$

**20.** *Use the estimated model to predict daily gas usage in a month when temperatures average 75°.*

**21.** *Does your estimate make sense to you? Why does the model give this result?*

## A Classic Example

Between 1595 and 1606 at the University of Padua, Galileo Galilei (1564–1642) conducted a series of famous experiments on the behavior of projectiles. Among these experiments were observations of a ball rolling down an inclined ramp (see diagram on facing page). Galileo varied the height at which the ball was released down the ramp, and then measured the horizontal distance which the ball traveled.

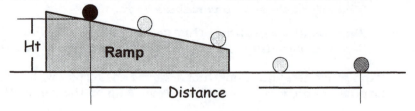

We'll begin by looking at the results of one of his experiments; the data are in **Galileo**. As you might expect, balls released at greater heights traveled longer distances. Galileo hoped to discover the underlying relationship between release height and horizontal distance. Both the heights and distances are recorded in *punti* (points; a unit of distance).

🖱 First, plot the data in the first two columns of the worksheet, with horizontal distance as the Y variable.

22. ***Does the graph suggest that distance and height are related? Is the relationship positive or negative?***

23. ***For what logical or physical reasons might we expect a non-linear relationship?***

Although the points in the graph don't quite fall in a straight line, let's perform a linear regression analysis for now.

🖱 Perform the regression, using **DistRamp** as the response.

24. ***Referring to your regression results, comment on the meaning and statistical significance of the slope and intercept, as well as the goodness of fit measures.***

25. ***Use the estimated regression equation to determine the release height at which a ball would travel 520 punti. (Note: think about how you can use your results to arrive at this value).***

As we did earlier, let's visualize the estimated regression line in relation to the sample points:

🖱 **Stat ➤ Regression ➤ Fitted Line Plot…** Select the appropriate variables.

26. ***Comment on the extent to which the line appears to "fit" the plotted points.***

27. ***Do you think your estimated release height for a 520 punti travel is probably accurate, or is it likely to be too high or too low? Explain.***

It should be clear that a linear model is not the very best choice for this set of data. Regression analysis is a very powerful technique, which is easily misapplied. In upcoming sessions, we'll see how we can refine our uses of regression analysis to deal with problems such as non-linearity, and to avoid abuses of the technique.

## *Moving On...*

Use the techniques and information in this session to answer the following questions. Explain or justify your conclusions with appropriate graphs or regression results.

### Galileo

1. Galileo repeated the rolling ball experiment with slightly different apparatus. In the second experiment, a horizontal shelf was attached to the lower end of the ramp. Use the data in the third and fourth columns of the worksheet to estimate the relationship between horizontal distance and release height.

2. At what release height would a ball travel 520 punti in this case?

### US

Investigate possible linear relationships between the following pairs of variables. In each case, comment on (a) why the variables might be related at all, (b) why the relationships might be linear, (c) the interpretation of the estimated slope and intercept, (d) the statistical significance of the model estimates, and (e) the goodness of fit of the model. (In each pair, the Y variable is listed first.)

3. Cars in use vs. population

4. Aggregate personal savings vs. aggregate personal income

5. Total federal receipts vs. aggregate personal income

6. GDP vs. aggregate civilian employment

### MFT

These are the Major Field Test scores, with student GPA and SAT results. Investigate possible linear relationships between students' MFT total scores and the following predictor variables. In each case, comment on (a) why the variables might be related at all, (b) why the relationships might be linear, (c) the interpretation of the estimated slope and intercept, (d) the statistical significance of the model estimates, and (e) the goodness of fit of the model. (In each question, the Y variable is the Total MFT score.)

7. GPA

8. Verbal SAT

9. Math SAT

## StateTrans

This worksheet contains transportation-related data for the United States.

10. Use linear regression analysis to investigate the relationship between the number of traffic fatalities in a state and the population of the state. Describe your findings and conclusions.

11. Instead of using population as the predictor variable, try using the number of motor vehicle registrations in the state. Compare these regression results with those in the prior question.

12. Now try using the number of drivers licenses issued by the state as the predictor variable, and compare the results with the two prior questions.

## Bodyfat

13. These are the body fat and other measurements of a sample of men. Our goal is to find a body measurement that can be used reliably to estimate body fat percentage. For each of the three measurements listed here, perform a regression analysis. Explain specifically what the *slope* of the estimated line means in the context of body fat percentage and the variable in question. Select the variable which you think is best to estimate body fat percentage.

    • Chest circumference
    • Abdomen circumference
    • Weight

14. Consider a man whose chest measurement is 95 cm, abdomen is 85 cm, and who weighs 158 pounds. Use your best regression equation to estimate this man's body fat percentage.

## Impeach

This worksheet contains the results of the Senate impeachment trial of President Clinton. Each senator could have cast 0, 1, or 2 guilty votes in the trial.

15. The worksheet contains a rating by the American Conservative Union for each senator. A very conservative senator would have a rating of 100. Run a regression using the number of guilty votes cast as the response (dependent) variable, and the rating as the predictor. Based on these results, would you say that political ideology was a good predictor of a senator's vote?

16. The file also includes the percentage of the vote cast for President Clinton in the senator's home state in the 1996 election. Run a regression to predict guilty votes based on this variable. Do your results suggest that electoral politics was a good predictor of a senator's votes?

17. Comment on the appearance of the scatterplots for these two regressions. Does linear regression analysis appear to be an ideal technique for analyzing these data? Explain.

## California

This worksheet contains some simple almanac data for the counties in California.

18. Perform, interpret, and compare regressions using the county population as the predictor, and the following demographic group sizes as the response variables:

- Number of persons aged 5 to 17
- Number of persons aged 18 to 64
- Number of persons aged 65 and older

19. We also have similar worksheets for the states of **Florida**, **Illinois**, **Michigan**, **New York**, **Ohio**, **Pennsylvania**, and **Texas**. Select two of these states, and perform the same analysis that you ran for California. Compare the regression results for the three states, and comment on why the results turn out as they do.

# Linear Regression (II)

## Objectives

In this session, you will learn to:

- Validate the assumptions for Least Squares regression by analyzing the *residuals* in a regression analysis
- Use a least squares line to estimate or predict $Y$ values
- Interpret the standard error of the regression

## Assumptions for Least Squares Regression

In the prior session we learned how to fit a line to a set of points. Minitab uses a common technique, called the *method of least squares*.[1] Though there are several other alternative methods available, least squares estimation is by far the most commonly used.

We can run a regression for any set of paired data values. However, if we intend to use our estimates for consequential decisions we must use the technique with care. The least squares method will yield unbiased, consistent, and efficient[2] estimates only when certain conditions are true. Recall that the basic linear regression model states that $X$ and $Y$ have a linear relationship, but that any observed $(x,y)$ pair will randomly deviate from the line. Algebraically, we can express this as:

$$y_i = \beta_0 + \beta_1 x_i + \varepsilon_i$$

---

[1] This method goes by several common names, but the term "least squares" always appears, referring to the criterion of minimizing the sum of squared deviations between the estimated line and the observed Y values.

[2] You may recall the terms *unbiased, consistent* and *efficient* from earlier in your course. This is a good time to review these definitions.

where:

> $x_i$, $y_i$ represent the $i^{th}$ observation of $x$ and $y$, respectively,
> $\beta_0$ is the intercept of the underlying linear relationship,
> $\beta_1$ is the slope of the underlying linear relationship, and
> $\varepsilon_i$ is the $i^{th}$ random disturbance, i.e., the deviation between the theoretical line and the observed value $(x_i, y_i)$

For least squares estimation to yield reliable inferences the following must be true about $\varepsilon$, the random disturbance.[3]

- *Normality*: At each possible value of $x$ the random disturbances are normally distributed; $\varepsilon|x_i$ follows a normal distribution.
- *Zero mean*: At each possible value of $x$ the mean of $\varepsilon|x_i$ equals 0.
- *Homoscedasticity*: At each possible value of $x$, the variance of $\varepsilon|x_i$ equals $\sigma^2$, which is constant.
- *Independence*: At each possible value of $x$, the value of $\varepsilon_i|x_i$ is independent of all other $\varepsilon_j|x_j$.

If these conditions are not satisfied and we use the least squares method, we run the risk that our inferences—the tests of significance and any confidence intervals we develop—will be misleading. Therefore, it is important to verify that we can reasonably assume that $x$ and $y$ have a linear relationship, and that the four above conditions hold true. The difficulty is that we cannot directly observe the random disturbances $\varepsilon_i$ because we don't actually know the location of the true regression line. We instead examine the *residuals*—the differences between our estimated regression line and the observed $Y$ values.

## *Examining Residuals to Check Assumptions*

By computing and examining the residuals, we can get some idea of the degree to which the above conditions apply in a given regression analysis. We will adopt slightly different analytic strategies depending on whether the sample data are cross-sectional or time series. Because the order of observations is at the heart of time series data we can investigate the independence assumption quite directly. With cross-sectional data we are often ignorant of the order in which observations were recorded, and have limited ability to check independence. We will start with a cross-sectional example.

---

[3] Some authors express these as assumptions about $y|x_i$.

🖱 **File ➤ Open Worksheet... StateTrans**. This is the worksheet with data about transportation in the 50 states in the U.S. Let's first check for a linear relationship between the number of cars registered in the state in 1998 and the population of the state as recorded in the 2000 Census.

🖱 **Graph ➤ Scatterplot...** Choose a simple scatterplot with regression. **Regist** is Y and **Pop** is X.

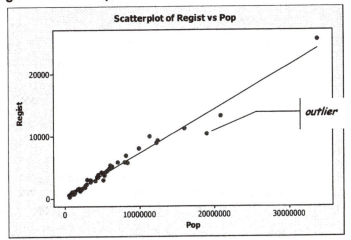

Based on this graph, it appears that there is an underlying linear relationship between the two variables. Next we perform the regression, and have Minitab compute and store the estimated y values and the residuals in the worksheet. Here is one several way to accomplish this.

🖱 **Stat ➤ Regression ➤ Regression...** Select **Regist** as the **Response** variable and **Pop** as the **Predictor**. Before clicking **OK**, click on **Graphs....** This button opens another dialog box (see next page), allowing you to specify various graphs for Minitab to create. Check **Four in one** and then click **OK**. Click **OK** in the Regression dialog.

This command generates four graphs of the residual values. Before learning to interpret the graphs, let's take a close look at the regression results themselves. We interpret the regression results exactly as in the prior session. This regression model looks quite good: the significance tests are impressive, and the coefficient of determination ($r^2$) is quite high.

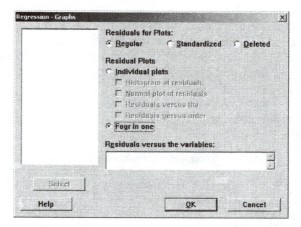

Before examining the residuals per se, note the lower section of the output, labeled "Unusual Observations." Four points are identified as unusual. The second has a very large negative residual. This state has a rather large population of 18,976,457 but relatively few cars registered for a state that size.

1. *This point is highlighted on the scatterplot on the previous page. Can you surmise the identity of the state? (Hint: on your screen, find the worksheet row number of that unusual observation). What might explain the relatively low number of registered vehicles in this state?*

All four unusual observations are marked on the right side of the output with an "R", denoting that they have large *standardized residuals*. This means roughly speaking that these points lie more than two standard errors vertically from the estimated line.

Two of the other unusual observations is marked with an "X," indicating an "observation whose X value gives it large influence." They are unusual because they exert "leverage" in the estimation of the slope. The two points in question are the right-most points in the graph. The outlying point farthest to the right represents California, and to its left is Texas. Because of their extreme X positions, the slope computation depends heavily on these two values; small differences in the Y value of either could have had substantial impact on the estimated slope.

In the Data Window, now scroll to the right. Notice that two new columns have appeared, labeled **RESI1** and **FITS1**, containing the estimated Y values for each state, and the residual for each. We can examine the residuals to help evaluate whether any of the four least squares assumptions should be questioned. We can inspect the residuals as surrogates for the random disturbances.

Recall that there are four assumptions. In this dataset the observations occur alphabetically by state and therefore we cannot really draw conclusions about the residuals being independent of one another. Of the other three assumptions, we are unable to use the residuals to evaluate the *zero mean* assumption, since the least squares method guarantees that the mean of the residuals will be zero.

However, we can examine the residuals to help decide if the other two assumptions—*normality* and *homoscedasticity*—are valid. The simplest tools for doing so are some graphs. The four residual graphs provide us with information about the assumptions.[4]

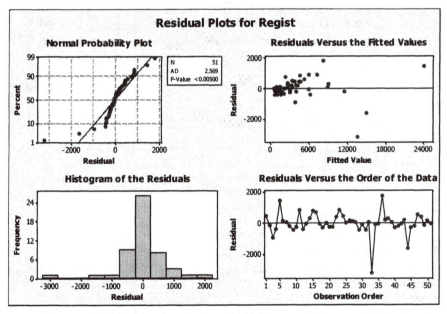

The two graphs on the left side relate to normality. For residuals that are normally distributed, the Normal Probability Plot will look like a 45° upward sloping diagonal line. The Histogram below it should look like a symmetrical, bell-shaped normal distribution. To the extent that the graphs deviate from these patterns, the normality assumption should

---

[4] Most standard statistics textbooks discuss the interpretation of residual graphs. Consult your text for a more detailed discussion.

be questioned.[5] In this case, the residuals do not appear to be normally distributed; at this point in our investigation we should not proceed to draw inferences based on this regression, since the normality assumption seems inappropriate. Notice the degree to which the large negative outlier affects these results.

The graph in the upper right is a plot of the residuals versus the fitted, or estimated, values[6]. This graph can give us insight into the assumption of equal variances (*homosceda*sticity), as well as the assumption that $X$ and $Y$ have a linear relationship. When both are true, the residuals will be randomly scattered in an even, horizontal band around a residual value of zero. Residuals that "fan out" from left to right, or right to left, signal *heteroscedasticity*, or unequal variances. A curved pattern suggests a non-linear relationship. Here, we see a "fan": pattern—the points are tightly clustered near the horizontal 0 line on the left of the graph, but notably spread out from the line as we look to the right. This suggests that we have unequal variances in the random scatter of the points. As with other techniques, a violation of the equal-variance assumption renders our inferences unreliable. Sessions 18 and 19 suggest some strategies for dealing with heteroscedasticity.

The graph in the lower right, the Residuals Versus the Order of the Data, is a sequential plot of the residuals, ordered by observation number. Mostly, this graph is relevant to time series data. Since the order of observations here is arbitrary (alphabetical by state name), we'll skip this graph for now, and discuss it in the next example.

## *A Time-Series Example*

Earlier, we noted that the assumption of independence is typically a concern in time series datasets. In particular, if the residual at time $t+1$ depends on the residual at time $t$, inference and estimation present some special challenges. Once again, our initial inclination will be to assume that the random disturbances *are* independent, and look for compelling evidence that they are not. As before, we do so by running a least squares regression, saving the fitted and residual values, and examining the residuals.

---

[5] For a more rigorous test of normality, note the Anderson-Darling Normality test reported in the probability plot. The null hypothesis is that the data are normally distributed; if the p-value is small, we would reject that hypothesis.

[6] Some authors prefer to plot residuals versus $X$ values; the graphs are equivalent.

Our next example uses time-series data, so that the sequence of these observations is meaningful. As in the prior session, we'll look at some annual data from the U.S. economy, and return to the relationship between aggregate personal savings and aggregate personal income. Open the file **US**. As in the last session, we'll run a regression with **PersSav** as the Y variable and **PersInc** as the X variable. Recall that the scatterplot for these two variables was not linear.

🖐 **Stat ➤ Regression ➤ Regression...** The response is **PersSav**, and the Predictor is **PersInc**. As in the prior regression, we want to examine the residuals. Since we have not changed our **Graphs** option from our prior regression command, we will once again get the four-in-one residual graph.

This time, because the sequence of residuals *is* meaningful, we should examine the sequential run chart of residuals in the lower right. What are we looking for? If the residuals are independent, then a negative residual at time $t$ should not affect the likelihood of a positive or negative residual at time $t+1$. Thus, if we see positive or negative residuals grouped sequentially, we might doubt this assumption.

2.   *In this chart, do you see areas where a consecutive series of residuals appear above or below the horizontal 0 (zero) line?*

Now look back at the other three residual plots.

3.   *What do you conclude about the assumptions of normality and homoscedasticity?*

## Issues in Forecasting and Prediction

One reason that we are concerned about the assumptions is that they affect the reliability of estimates or forecasts. To see how we can use Minitab to make and evaluate such forecasts, we'll turn to another example. Open the file called **Utility**.

This file contains time-series data about the consumption of natural gas and electricity in my home. Also included is the mean monthly temperature for each month in the sample. As in the prior session, we'll model the relationship between gas consumption and temperature, and then go on to forecast gas usage in a month when mean temp is 32 degrees.

As before, we will "plug" 32 degrees into our estimated regression model to get a point estimate. However, we can also develop either a Confidence or a Prediction interval, depending on whether we wish to estimate the *mean* gas use in all months averaging 32 degrees, or the actual gas use in one particular month averaging 32 degrees.

> ⌐ **Stat ➤ Regression ➤ Regression…** As before, **GaspDay** is the response, **MeanTemp** is the predictor. Before clicking OK, we want to select **Options…**.

> > ⌐ **Options…** In the options dialog, type 32 in the box marked **Prediction intervals for new observations**, to generate confidence and prediction intervals. Click **OK**.

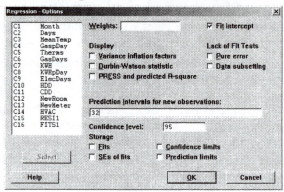

Look at the regression results in the Session Window. Below the unusual observations, you'll see some new output displaying a predicted values of **GaspDay**, including 95% confidence and prediction intervals.

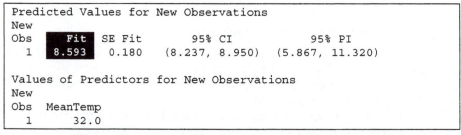

```
Predicted Values for New Observations
New
Obs    Fit    SE Fit      95% CI            95% PI
  1   8.593   0.180   (8.237, 8.950)   (5.867, 11.320)

Values of Predictors for New Observations
New
Obs   MeanTemp
  1      32.0
```

> **4.  What do these intervals tell you about natural gas usage when the average daily temperature is 32 degrees?**

Before relying too heavily on this estimate, let's look at the residuals. Remember that these are time-series data.

> **5.  Do you see any evidence that this model might violate any of our regression assumptions? Explain.**

It is important to think about the ways in which such violations can affect estimation. To get a better look at the problem here, let's re-plot the data, superimposing our estimated regression line.

🖱 **Stat ▶ Regression ▶ Fitted Line Plot...** In this dialog, enter **GaspDay** for the response variable and **MeanTemp** for the predictor.

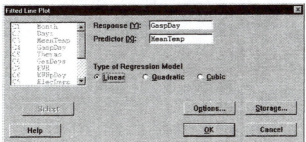

Along the X axis, visually locate 32 degrees, and look up at the regression line and at the observed data points. Note that most of the observed data at such low temperatures lies *above* the regression line, and that the points gently arc around the line. Therefore, the estimated value for 32° is probably too low. What can we do about that? We'll see some solutions in a future session.

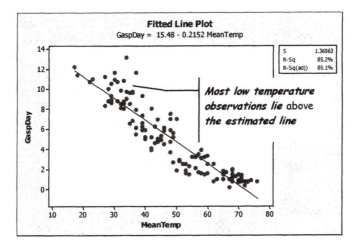

## A Caveat about Atheoretical Regression.

Linear regression is a very powerful tool that has numerous uses. Like any tool, it must be used with care and thought, though. To see how thoughtless uses can lead to bizarre results, try the following.

🖰 **File ➤ Open Worksheet...** Open the **Anscombe** worksheet file.

This dataset contains eight columns, representing four sets of X-Y pairs. We want to perform four regressions using X1 and Y1, X2 and Y2, etc. By now, you should be able to run four separate regressions.

**6.   *After you do so, look closely at the four sets of results, and comment on what you see.***

Based on the regressions, it is tempting to conclude that the four x-y pairs all share the same identical relationship. Or is it?

🖰 **Graph ➤ Scatterplot...** Construct four scatter plots (Y1 vs. X1, Y2 vs. X2, etc.)  What do you see? Remember that each of these four plots led to the four virtually identical regressions.

## Moving On...

Using the techniques of this session, perform and evaluate regressions *and* residual analyses to investigate the following relationships.

## US

Use your regressions and Minitab to predict the dependent variable as specified in the question. In each instance, report the estimated regression equation, and explain the meaning of the slope and intercept.

1.   Cars in use vs. Population (predict when **Pop** = 245,000 i.e., 245 million people)

2.   **FedRecpt** vs. **PersInc** (predict when **PersInc** = 5000 i.e., 5 trillion dollars)

## StateTrans

*Note*: The "unusual" state from the first example is New York. In the worksheet, enter an asterisk (missing data) in the **Pop** column for New York. Now re-do the first example from this session, omitting the unusual case of New York.

3. How, specifically, does this affect (a) the regression and (b) the residuals? Compare the slope and intercepts of your two regressions, and comment on what you find.

## Bodyfat

4. In Session 16, you ran some regressions using these data. This time, perform a linear regression and residual analysis using **FatPerc** as the response variable, and **Weight** as the predictor. Estimate a fitted value for a man who weighs 158 pounds.

5. Do these sample data suggest that the least squares assumptions have been satisfied in this case?

6. What is the 95 percent interval estimate of mean body fat percentage among all men who weigh 158 pounds?

7. What is the 95 percent interval estimate of body fat percentage for one *particular* man who weighs 158 pound?

## Galileo

8. In the previous session, we noted that horizontal distance and release height (first two columns) did not appear to have a linear relationship. Re-run the regression of distance (Y) versus height, and construct the residual plots. Where in these plots do you see evidence of non-linearity?

9. Repeat the same with columns 3 and 4. Is there any problem with linearity here? Explain.

## MFT

In the prior session, you ran one or more regression models with Total MFT Score as the dependent variable. Repeat those analyses, evaluating the residuals for each model.

10. GPA

11. Verbal SAT

12. Math SAT

## Impeach

13. Perform a regression and residual analysis with number of guilty votes as the dependent variable and conservatism rating as the independent variable. Discuss unusual features of these residuals. To what extent do the assumptions seem to be satisfied?

14. Perform a regression and residual analysis with number of guilty votes as the dependent variable and percentage of the vote received by President Clinton in 1996 as the independent variable. Discuss unusual features of these residuals. To what extent do the assumptions seem to be satisfied?

## California

15. Perform a regression and residual analysis using the county population as the predictor, and the following demographic group sizes as the response variables:
    - Number of persons aged 5 to 17
    - Number of persons aged 18 to 64
    - Number of persons aged 65 and older

16. Use a linear regression model to estimate the 18 to 64 year old population of a county with a total population of 150,000 people. Comment on how accurate you think this estimate is.

17. We also have similar worksheets for the states of **Florida**, **Illinois**, **Michigan**, **New York**, **Ohio**, **Pennsylvania**, and **Texas**. Select two of these states, and create a regression model to estimate the 18 to 64 year old population of a county with a total population of 150,000 people, just as you did for California. Compare the regression, residual, and estimation results for the three states, and comment on why the results turn out as they do.

## Florida Votes

During the 2000 Presidential Election, there were a number of controversial occurrences in Florida. Among the controversies was the use of a "butterfly" ballot in Palm Beach County, which some voters claim to have found confusing. Specifically, the news media reported that voters intending to cast their votes for Al Gore may instead have selected Pat Buchanan.

18. Perform a regression and residual analysis using the number of votes cast for Buchanan as the response and the number cast for Bush as the predictor. Comment on what you find in this regression, noting especially the fitted value for Palm Beach County as reported among the unusual observations.

19. Complete a similar analysis using the number of votes cast for Buchanan as the response and the number cast for Gore as the predictor. Comment on what you find in this regression, noting especially the fitted value for Palm Beach County as reported among the unusual observations.

20. Ultimately President Bush carried the entire state of Florida by 547 votes over Vice President Gore. Based on your analysis, comment on the relative size of this margin of victory.

# Session 18

# Multiple Regression

### *Objectives*

In this session, you will learn to:

- Improve a regression model using multiple regression analysis
- Interpret multiple regression coefficients
- Incorporate qualitative data into a regression model
- Diagnose and deal with multicollinearity

### *Going Beyond a Single Explanatory Variable*

In our previous sessions using simple regression we examined several bivariate relationships. In some examples we found a statistically significant relationship between the two variables, but also noted that much of the variation in the dependent variable remained *unexplained* by a single independent variable, and that the standard error of the estimate was often rather high compared to the standard deviation of the dependent variable.

For many dependent variables we may identify numerous possible causes. The statistical tool of *multiple regression* enables us to identify those variables simultaneously associated with a dependent variable and to estimate the separate and distinct influence of each variable on the dependent variable.

For example, recall our Session 16 analysis of aggregate personal savings in the United States as a function of personal income. In that case, we found a distinctly non-linear relationship with a coefficient of determination equal to only about 59%. Despite steady increases in aggregate personal income, aggregate personal savings in the United States has been declining since the early 1990's. There are probably several reasons behind the decline in savings, but we might hypothesize that the rapid increase in common stock values and the strong growth of the U.S. economy might have increased consumption and depressed savings. In our **US** dataset, we have a variable called **Dow**, and it is the Dow-Jones Industrial Average on January 1 of each year. The Dow index does not represent all stocks, but it is a widely-followed indicator of the general state of the markets. As such, it is a reasonable *surrogate variable* to use in this example. Let's see if **Dow** and **Savings** are related. Open **US**. Since we are interested in relationships among three variables, a *matrix plot* is a good tool to use.

🖱 **Graph ➤ Matrix Plot...**  Choose a simple plot, and specify the variables **PersSav, PersInc,** and **Dow**.

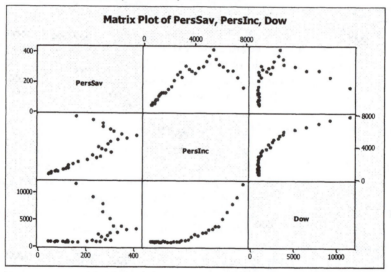

In the resulting plot, we see scatter plots relating each pairing of these three variables. In the first row, both of the graphs have **PersSav** on the Y-axis; in the first column, **PersSav** forms the X axis. You should recognize the plot of Savings vs. Income.

1.   *What do you see in the plot of savings and the Dow-Jones average? Does this appear to be a linear relationship?*

We can also think about these pairings in terms of the correlations between the variables. To compute the correlation coefficients for each pair, do the following:

🖱 **Stat ➤ Basic Statistics ➤ Correlation...** Here again, specify the variables **PersSav**, **PersInc**, and **Dow**.

In the session window, you will find the *correlation matrix*, which reports the correlation between each pairing of the three variables. The matrix is a triangular arrangement of columns and rows; the correlation coefficient for a pair of variables appears at the intersection of the corresponding row and column. For example, savings and income have a correlation of 0.769, while savings and the Dow average have a correlation of only 0.348. Both correlations are positive, suggesting (for example) that income and aggregate savings rise and fall together, though imperfectly.

```
Correlations: PersSav, PersInc, Dow

            PersSav   PersInc
PersInc      0.769
             0.000

Dow          0.348     0.867
             0.044     0.000
```

It is important to recognize that these correlations and the matrix plot refer only to the pair of variables in question, without regard to the influences of other variables. Above though we theorized a *multivariate relationship*: savings varies *simultaneously* with income and stock valuations. That is to say, we suspect that stock market prices affect savings in the context of a given income level. Therefore we can't merely look at the relationship of interest rates and savings without taking income into account. Multiple regression allows us to do just that. Let's see how.

In simple linear regression, we visualize a relationship as a straight line. In this model, we need to think of a regression *plane* in three-dimensional space.

To help visualize what the relationship might look like, we need to add a dimension to our scatter plot. Minitab lets us do that as follows:

🖱 **Graph ➤ 3-D Scatterplot...** Choose a simple scatter plot, and complete the dialog box as shown here:

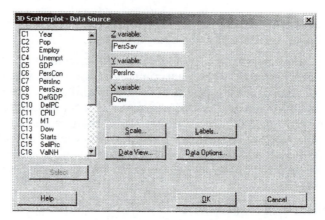

The scatter plot displays a three-dimensional graph of the data, shown from one perspective. As we view the data from this particular perspective, the observed points seem for form a curve that rises and then falls. If we change our vantage point, though, a different picture emerges. To shift our perspective, we need to add a specialized tool. For this reason, when this graph opens, you should find that a new menu bar pops up near the graph window.[1]

🖑 Among the new set of buttons, locate the button with a clockwise pointing arrow marked **z**. Click on this button several times, and then hold the mouse button for a few moments.

*Click on this button to rotate the graph clockwise around the Z-axis.*

As you click on the rotation control button, your graph will rotate in space revealing a "necklace" of points. Continue to observe the graph through several rotations.

2. *Describe what you notice as you spin the graph in this manner.*

3. *Find one perspective from which the points appear to lie on a single plane, and print the graph. Describe what you see on your screen.*

---

[1] If the 3-D Graph Tools menu bar does not open, do the following. Move your cursor to the menu bar at the top of the screen, and right click. This will open a list of available menu bars. Check 3-D Graph Tools.

We can estimate the coefficients in this multivariate model using the regression procedure, much as before. Algebraically, our model in this example is this:

$$Savings_i = \beta_0 + \beta_1 Income_i + \beta_2 Dow_i + \varepsilon_i$$

Stat ➤ Regression ➤ Regression...  Select **PersSav** as the Response variable, and both **PersInc** and **Dow** as Predictors. Also select the four-in-one residual graph. The upper portion of the results is shown here:

```
Regression Analysis: PersSav versus PersInc, Dow

The regression equation is
PersSav = 39.5 + 0.0882 PersInc - 0.0502 Dow

34 cases used, 2 cases contain missing values

Predictor         Coef    SE Coef        T       P
Constant        39.507      7.333     5.39   0.000
PersInc       0.088196   0.003461    25.48   0.000
Dow          -0.050162   0.002905   -17.27   0.000

S = 22.0658    R-Sq = 96.0%    R-Sq(adj) = 95.7%
```

Look in the Session Window for the regression results. They should look quite familiar, with the only differences being an additional line in the table of coefficients, corresponding to **Dow**, and some new output below the ANOVA table. Look at the coefficients.

We now have one intercept (Constant) and two slopes, one for each of the two explanatory variables. The intercept represents the value of Y when *all* of the X variables are equal to zero. Each slope represents the marginal change in Y associated with a one-unit change in the corresponding X variable, *if the other X variable remains unchanged.* For example, if personal income were to increase by one billion dollars, and mortgage rates were to remain constant, then savings would increase by .0088196 billion dollars ($88 million). Look at the coefficient for **Dow**.

4.   *What, specifically, does this coefficient tell us?*

## Significance Testing and Goodness of Fit

In linear regression, we tested for a significant relationship by looking at the *t*- or *F*-ratios. In multiple regression, the two ratios test two different hypotheses. As before, the *t*-test is used to determine if a slope equals zero. Thus, in this case, we have two tests to perform:

| Persinc | Dow |
|---------|-----|
| $H_0: \beta_1 = 0$ | $H_0: \beta_2 = 0$ |
| $H_A: \beta_1 \neq 0$ | $H_A: \beta_2 \neq 0$ |

The *t*-ratio and *P*-value in each row of the table of coefficients tell us whether or not to reject each of the null hypotheses. In this instance at the 0.05 level of significance, we reject in both cases, due to the very low p-values. That is to say both **Persinc** and **Dow** have statistically significant relationships to **PersSav**.

The *F*-ratio in a multiple regression is used to test the null hypothesis that all of the slopes are equal to zero:

$$H_0: \beta_1 = \beta_2 = 0 \quad \text{vs.} \quad H_A: H_0 \text{ is not true.}$$

Note that the alternative hypothesis is different from saying that all of the slopes are non-zero. If one slope were zero and the other were not, we would reject the null in the *F*-test. In the two *t*-tests, we would reject the null in one, but fail to reject it in the other.

Finally, let's consult $r^2$, the Coefficient of Multiple Determination. Adding any *X* variable to a regression model will tend to inflate $r^2$. To compensate for that inflation, we adjust the coefficient of determination to account for both the number of *x* variables in the model, and for the sample size.[2] In multiple regression analysis, we want to consult the *adjusted* $r^2$ figure. In this instance, the addition of another variable really does help to explain the variation in *Y*.

In this regression, the adjusted $r^2$ equals 95.7%; in the prior session, using only income as the predictor variable, the adjusted $r^2$ was only 58%. We would say that, by including the Dow in the equation, we are accounting for an additional 37.7% (95.7 – 58) of the total variation in aggregate personal savings.

### *Residual Analysis and Prediction*

As in simple regression, we want to evaluate our models in terms of their predictive performance and the degree to which they conform to the assumptions concerning the random disturbance terms. Look at the residual graphs from your regression.

---

[2] See your textbook for the formula for adjusted $r^2$. Note the presence of *n* and *k* (the number of predictors) in the adjustment.

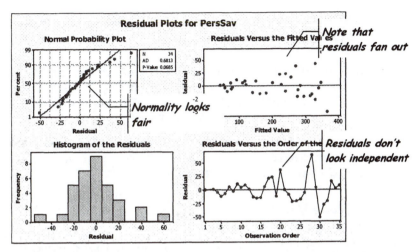

We interpret these residual plots exactly as we did before. In these particular graphs, you should note that there is doubt about both homoscedasticity and independence of the random errors. Besides explaining more variation than a simple regression model, a multiple regression model can sometimes resolve a violation of one of the regression assumptions. This is because the simple model may assign too much of the unexplained variation to $\varepsilon$, when it should be attributed to another variable. As such, the residuals are not genuinely random.

Even though these residuals indicate some likely violations of the least squares assumptions, let's continue with the example to illustrate how we would make an estimate using a multiple regression model. Let's see how well the model performs in estimating savings for the year 1999, when U.S. aggregate income was 7791.2 billion dollars and the Dow was at 11497.1.

🖰 **Stat ➤ Regression ➤ Regression…**We'll leave our variable selection as before, but in the **Options** dialog, enter the values **7791.2** and **11497.1** into the **Prediction intervals for new observations** box.

💻 NOTE: In the **Prediction intervals for new observations** box of the **Options** dialog, you must enter the predictor values in the same order as the Predictor variables in the Regression dialog. That's how Minitab 'knows' that 7791.2 is the value for Income, not Dow.

The regression output is virtually the same as before (it's the same regression), but now the prediction for 1999 appears at the end of the output.

5.   **What is the predicted, or fitted, value for 1999? The actual figure for aggregate savings in 1999 was 158.3 billion dollars. How closely did the model estimate aggregate savings for 1999?**

6.   **As before, note that the output includes both a 95% confidence interval and a 95% prediction interval. What do these intervals tell you?**

## Adding More Variables

This model improved the previous simple regression. Let's see if we can do better still. Suppose we hypothesize that during times of high unemployment, people are reluctant to spend too much of their income, and therefore have an extraordinary incentive to save, after controlling for income and stock market effects. In other words, we want to add the Unemployment Rate variable (**Unemprt**) to the equation. Moreover, if this theory is right, we'd expect the coefficient of **Unemprt** to be positive.

⌐🖰  **Stat ➤ Regression ➤ Regression…**  Add **Unemprt** to the list of predictors. Before clicking **OK**, click on **Options** and delete the predictor values. Then perform the regression.

Compare this regression output to the earlier results.

7.   **What was the effect of adding this new variable to the model? Obviously, we have an additional coefficient and t-ratio. Assuming a significance level of $\alpha = 0.10$, does that t-ratio indicate that** Unemprt **has a significant relationship to aggregate savings, when we control for income and the Dow? Does the coefficient have a positive sign, as expected?**

8.   **What else changed? Look, in particular, at the adjusted $r^2$, the ANOVA, and the values of the previously estimated coefficients. Can you explain the differences you see?**

The addition of a new variable can also have an impact on the residuals. In general, each new model will have new residual plot.

9.   **Examine these residual graphs, and see what you think. Do the least squares assumptions appear to be satisfied?**

## *A New Concern*

In a model including income, the Dow-Jones Industrials average, and unemployment rates, we have accounted for about 96% of the variation in Aggregate Personal Savings, and have a standard error of the regression of about 21 billion dollars. This is quite an improvement over the simple model in which $r^2$ was 56% and $s$ was $70 billion. Suppose we wanted to try to improve the model even further, and hypothesize that at times of high inflation, people might save even more than usual, with consumption unattractive due to high prices. Let's add the Consumer Price Index (**CPIU**) to the model, and re-run the regression with C.P.I. added to the list of predictors.

Look at this regression output and take special note of total variation explained, the standard error $s$, and of the other regression coefficients and test statistics. Only the estimated slope for the Dow appears to be statistically significant! What is happening here?

This is an illustration of a special concern in multiple regression: *multicollinearity*. When two or more of the predictor variables are highly correlated in the sample, the regression procedure cannot determine which of the predictor variables concerned is associated with changes in $Y$. In a real sense, regression cannot disentangle the individual effects of each $X$. In this instance, the culprits are C.P.I. and Personal Income.

⌐🖰 **Stat ➤ Basic Statistics ➤ Correlation...** Select all five variables for this correlation matrix : **PersSav PersInc Dow Unemprt CPIU**.

Note that C.P.I. and Savings are highly correlated (0.852), which ordinarily is good; however C.P.I. and Personal Income have a nearly perfect correlation of 0.987. This is the root of the problem, and with this sample, can only be resolved by eliminating one of the two variables from the model. Which should we eliminate? We should be guided both by theoretical and numerical concerns: we have a very strong theoretical reason to believe that income belongs in the model, but C.P.I. has a stronger numerical association with savings. Given a strong theoretical case, it is probably wiser to retain income and omit the C.P.I.[3]

In assessing the quality of a multiple regression model, we always want to think about the same issues:

- Do the variables make theoretical sense?
- Do the estimated coefficients have the expected signs?

---

[3] We could keep income, and use the annual *change* in the C.P.I., which measures inflation in the given year. The idea here is to illustrate the effects of multicollinearity, and to emphasize that the problem should not be ignored.

- Are the estimated coefficients statistically significant?
- How good is the fit, as measured by the adjusted coefficient of determination?
- How small is the standard error in comparison to the standard deviation of the response variable?
- Are there many unusual observations?
- Do the residuals indicate violations of the Least Squares assumptions?

## *Working with Qualitative Variables*

Multiple regression analysis is the appropriate tool when we believe that a quantitative response variable has multiple causes. Thus far our models have included only quantitative predictor variables. What happens when one of the causes is qualitative? Open the file called **Utility**, which you have seen before. In our earlier regression sessions, you created a simple linear model to estimate natural gas consumption using mean monthly temperature as the predictor.

During the study period we enlarged the house by adding a room and thus increased our need for heat during the winter months. We might have measured the volume of the house pre- and post-addition, but our dataset does not include such a measurement. We do however have a variable that indicates when the room was added. The distinction between the "old, smaller" house and the "new, larger" house is of course a categorical variable. All of the variables we have used in regression analysis so far are quantitative. It is possible to include a qualitative predictor variable in a regression analysis if we come up with a way to represent a categorical variable numerically.

We do this with a fairly simple trick known as a *dummy variable*. A dummy variable is an artificial binary variable that assumes arbitrarily chosen values to represent two distinct categories.[4] Ordinarily, we set up a new variable that equals 0 for one category, and 1 for the other. In this dataset, we have a variable called **NewRoom** which happens to equal 0 for months prior to the addition and 1 for months after the room was added. We can use that variable in this case.

Before running a regression, let's look at the data.

🖰 **Graph ➤ Scatterplot...** Choose With Regression and Groups, as shown below, left.

----

[4] It is possible to represent multinomial variables using several dummy variables. Consult your textbook or instructor.

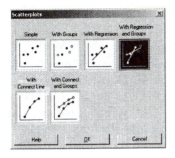

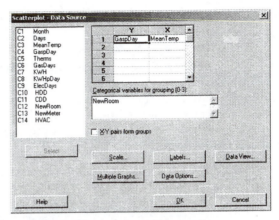

🖱 In the scatter plot dialog (above, right) select **GaspDay** as the Y variable, **MeanTemp** as the X variable, and **NewRoom** as the Categorical variable. Click **OK**.

Now look closely at the resulting scatter plot, shown below. We have all the data that we saw in Session 16, but here there are two new elements. The pre- and post-construction points are differentiated by the shape of the symbols. The squares represent observations made since the room was added, and the circles are pre-construction data. We also see two regression lines, corresponding to the two configurations of the house. In other words, by accounting for this qualitative variable we are creating two simple linear models, each with a different slope and intercept.

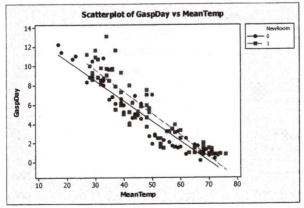

We will incorporate the dummy variable into our model as a separate variable and as an *interaction term* in the equation. In this way, the dummy will potentially affect both the slope and the intercept of the regression lines. Our model will be;

$$GaspDay_i = \beta_0 + \beta_1 MeanTemp + \beta_2(NewRoom \times MeanTemp) + \beta_3 NewRoom$$

Before the new room was added, **NewRoom** equals 0. Thus, for those observations, the final two terms in the equation vanish. When **NewRoom** equals 1, the final two terms have the effect of changing the slope and intercept of the line. Let's see how this works. First, we need to create a new variable that represents the interaction of the temperature and new room variables.

 **Calc ➤ Calculator...** Create a new variable called **TempxRoom**, equal to **MeanTemp * NewRoom**, as shown here.

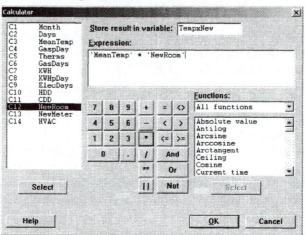

 Now run the regression with **GaspDay** as the **Response**, and **MeanTemp**, **TempxNew**, and **NewRoom** as the **Predictors**.

Now look at the regression results in the Session Window. We find that the coefficient of the interaction term is not statistically significant, but the other coefficients are. This indicates that the new room did change the intercept of the equation but did not change the slope substantially. Therefore, we should re-run the regression analysis without the interaction term.

 Run the regression again, this time deleting **TempxNew** from the list of predictors.

Look closely at the results and at the regression equation itself. Think for a moment about the meaning of the coefficient of the **NewRoom** variable. With the original size of the house, **NewRoom** = 0, and this equation becomes:

```
GaspDay = 15.1 - 0.217 MeanTemp + .802(0)
        = 15.1 - 0.217 MeanTemp
```

On the other hand, once the room was added the equation is:

```
GaspDay = 15.1 - 0.217 MeanTemp + .802(1)
        = 15.9 - 0.217 MeanTemp
```

In other words, we are looking at two parallel lines whose intercepts differ by .802 therms per day. The impact of the dummy variable, introduced into the equation in this way, is to alter the *intercept* for the two different categories.

**10.** ***Now that we know what the estimated equation is, let's go on to evaluate this particular model, as we did above. In evaluating the model, consider the issues cited on page 243.***

## Moving On...

### Utility

1. The worksheet also contains data about electricity usage in the home. The response variable is kilowatt-hours per day (**KWHpDay**). Create a multiple regression model using any of these variables;

   - **MeanTemp**
   - **HDD** (Heating degree-days)
   - **CDD** (Cooling degree-days)
   - **NewRoom**
   - **HVAC** (Dummy variable indicating when we renovated the heating system and added air conditioning)

   You may choose any one you wish, so long as you can explain how it logically might affect electricity usage once the other variables in your model are accounted for. Then run the regression model including the variables you have chosen, and evaluate the regression as we have done in this session. In your evaluation, you should comment on these questions:

- Are signs of coefficients correct?
- Are the relationships statistically significant?
- Do the residuals suggest that the assumptions are satisfied?
- Is there any evidence of a problem with multicollinearity?
- How much of the variation in monthly electricity consumption does your model account for?

## StateTrans

2. Develop a multiple regression model to estimate the number of fatalities using as many of the other available variables as you see fit (except **FatCrash**). One variable you might want to investigate is the Blood Alcohol Content threshold (**BAC**) which represents the legal definition of driving while intoxicated in each state. Do states which permit a higher threshold have more fatalities, other things being equal?

## Bodyfat

3. Develop a multiple regression model to estimate the body fat percentage of an adult male (**FatPerc**), based on one or more of the following easily-measured quantities:

- Age (years)
- Weight (pounds)
- Abdomen circumference (in cm)
- Chest circumference (in cm)
- Thigh circumference (in cm)
- Wrist circumference (in cm)

You should refer to a matrix plot or correlation matrix (or both) to help select variables. Your model may contain any or all of the listed predictor variables. The model that you select must (a) make sense, (b) have good significance test results, (c) have acceptable residuals (d) as high an $r^2$ as you can get, and (e) have no multicollinearity. Also discuss possible logical problems with using a linear model to estimate body fat percentage.

## Sleep

4. Develop a multiple regression model to estimate the total amount of sleep (**Sleep**) required by a mammal species, based on one or more of the following variables:

   - Body weight
   - Brain weight
   - Lifespan
   - Gestation

   You should refer to a matrix plot or correlation matrix (or both) to help select variables. Your model may contain any or all of the listed predictor variables. The model that you select must (a) make sense, (b) have good significance test results, (c) have acceptable residuals (d) as high an $r^2$ as you can get, and (e) have no multicollinearity. Also discuss possible logical problems with using a linear model to estimate sleep requirements.

## Labor2

5. Develop a multiple regression model to estimate the mean number of new weekly unemployment insurance claims (**Claims**). Select predictor variables from the other variables in the worksheet, based on theory, and the impact of each variable in the model.

   You should refer to a matrix plot or correlation matrix (or both) to help select variables. Your model may contain one, two, three, or all four predictor variable, aiming for a model which (a) makes sense, (b) has good significance test results, (c) has acceptable residuals (d) as high an $r^2$ as you can get, and (e) has no multicollinearity.

## Impeach

6. Develop a multiple regression model to estimate the number of guilty votes cast by a senator, using as many available variables as you see fit. As always, be sure you can explain why each independent variable is in the model.

### US

7. Develop a multiple regression model to estimate the annual number of housing starts, using any predictor variables available to you in the worksheet. As always, provide a logical theoretical reason for incorporating each variable, and evaluate your model using the same criteria described in the session.

### Mft

8. Create a 3-D scatter plot with student Major Field Test scores on the Z axis, and their verbal and math SAT scores on the X and Y axes. Manipulate the scatter plot to decide whether there appears to be a regression plane.

9. Run the multiple regression analysis to determine whether MFT scores can be accurately predicted knowing students' SAT scores. Discuss your findings.

### Student

10. Develop a multiple regression model for student weight using the available variables in the worksheet (except ideal weight). As always, provide a logical theoretical reason for incorporating each variable, and evaluate your model using the same criteria described in the session.

# Non-Linear Models

## *Objectives*

In this session, you will learn to:
- Improve a regression model by transforming the original data
- Interpret coefficients and estimates using transformed data

## *When Relationships are not Linear*

In our regression models thus far, we have assumed *linearity*; that is, that Y changes by a fixed amount whenever an X changes by one unit, other things being equal. The linear model is a good approximation in a great many cases. However, we also know that some relationships probably are not linear. Consider the "law of diminishing returns" as illustrated by a weight-loss diet. At first, as you reduce your calories, pounds may fall off quickly. As your weight goes down, though, the rate at which the pounds fall off may diminish.

In such a case, X and Y (calories and weight loss) are indeed related, but *not in a linear fashion*. This session provides some techniques that we can use to fit a *curve* to a set of points. The basic strategy in each case is the same. We attempt to find a function whose characteristic shape approximates the curve of the points. Then, we'll apply that function to one or more of the variables in our worksheet, until we have two variables with a generally linear relationship. Finally, we'll perform a linear regression using the transformed data. We begin with an artificial example.

## *A Simple Example*

We begin with a very familiar non-linear relationship between two variables, in which Y varies with the square of X. The formal model (known as a quadratic model) might look like this:

$$Y = 3X^2 + 7 + \varepsilon$$

First, let's create a set of data that reflects an exact relationship.

🖰 Label the first three columns of a blank Minitab worksheet as **X**, **Xsqr**, and **Y**.

🖰 **Calc ➤ Make Patterned Data ➤ Simple Set of Numbers...** Indicate that you want to store the data in **C1 (X)**, that the first value is **1**, and the last value is **20**. Then click **OK**. This will fill the X column with the values 1 through 20.

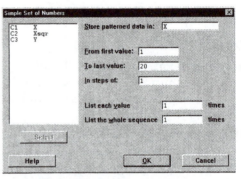

🖰 **Calc ➤ Calculator**. In **Xsqr**, calculate the square of X.

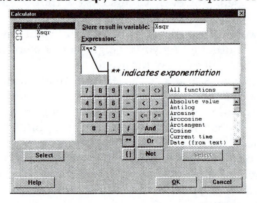

🖰 Now, set Y equal to 3*X² + 7. Bring up the calculator again, storing the result in **Y**, and the expression should read:

**3*Xsqr + 7**

🖱 Click **OK** on the calculator, and look at the Data Window. You will see three columns of numbers. The first row is 1, 1, 10 and the second is 2, 4, 19 (see first few lines below).

| | C1 | C2 | C3 | C4 |
|---|---|---|---|---|
| | X | Xsqr | Y | |
| 1 | 1 | 1 | 10 | |
| 2 | 2 | 4 | 19 | |
| 3 | 3 | 9 | 34 | |
| 4 | 4 | 16 | 55 | |
| 5 | 5 | 25 | 82 | |
| 6 | 6 | 36 | 115 | |
| 7 | 7 | 49 | 154 | |
| 8 | 8 | 64 | 199 | |
| 9 | 9 | 81 | 250 | |
| 10 | 10 | 100 | 307 | |

Worksheet 1 ***

If we plot Y versus X, and then plot Y versus Xsqr, the graphs look like this:

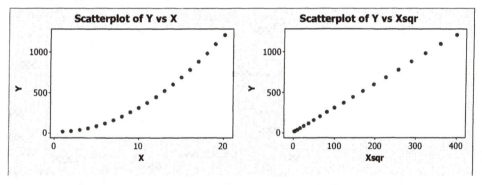

In the left graph, we see the distinct parabolic shape of a quadratic function. $Y$ increases each time that $X$ increases, but does so by an *increasing* amount. Clearly, $X$ and $Y$ are related, but just as clearly, the relationship is *curvilinear*. In the second graph, we have a perfect straight line that is an excellent candidate for simple linear regression.

1. *If you were to run a regression of Y on Xsqr, what would the slope and intercept be? (Do it, and check yourself.)*

This illustrates the strategy we noted above: When we have a curved relationship, we'll try to find a way to transform one or more of our variables until we have a graph that looks linear. Then we can apply our familiar and powerful tool of linear regression to the *transformed* variables. *As long as we can transform one or more variable and retain the basic functional form of y as a sum of coefficients times variables*, we can use linear regression to fit a curve. That is the basic idea underlying the next few examples. Minitab provides several ways to approach such examples, and we will explore two of them.

## Some Common Transformations

In our first artificial example, we squared a single explanatory variable. As you may recall from your algebra and calculus courses, there are many common curvilinear functions, such as cubics, logarithms, and exponentials. In this session, we'll use a few of the many possible transformations, just to get an idea of how one might create a mathematical model of a real world relationship.[1]

Let's begin with an example from Session 16 (and about 400 years ago): Galileo's experiments with rolling balls. Recall that the first set of data plotted out a distinct curve:

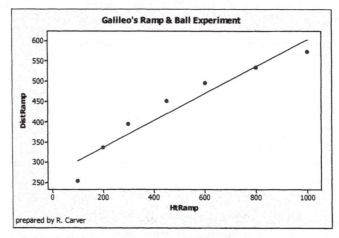

Note that the ball rolled further when started from a greater height, but that the increased horizontal roll diminishes as heights increase. A straight line is not a bad fit, but we can do better with a different functional form. In fact, Galileo puzzled over this problem for

---

[1] Selection of an appropriate model should be theory-driven, but sometimes can involve trial and error. The discussion of theoretical considerations is beyond the scope of this book; consult your primary text and instructor for more information.

quite some time, until he eventually reasoned that horizontal distance might vary with the square root of height.[2] If you visualize the graph of $y = +\sqrt{x}$ , it looks a good deal like this scatterplot: it rises quickly from left to right, gradually becoming flatter.

🖱 Open the worksheet called **Galileo**, and use the **Calculator** to create a new variable equal to the *square root* of **HtRamp**. Label that variable **SqrtHt.**

🖱 **Graph ➤ Scatterplot…** Put **DistRamp** is on the Y axis, and **SqrtHt** on the X axis, generating this graph:

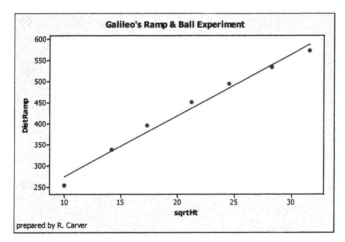

While not as dramatic as our first dataset, we see that this transformation does tend to straighten out the curved line. What is more, it makes theoretical sense that friction will tend to "dampen" the horizontal motion of a ball rolled from differing heights, just as the square root function represents. Other functional transformations may align the points more completely, but don't make as much sense, as we'll see later.

Now that we've straightened the dots, what do we do next? First, we run a regression, with **DistRamp** as the response variable, and **SqrtHt** as the lone predictor variable. Use the regression command to generate these results:

---

[2] For an interesting account of his work on these experiments, see Dickey, David A. and J. Tim Arnold, "Teaching Statistics with Data of Historic Significance." *Journal of Statistics Education,* v. 3, no. 1, 1995.

---

**Regression Analysis: DistRamp versus SqrtHt**

```
The regression equation is
DistRamp = 129 + 14.5 SqrtHt

Predictor           Coef        StDev           T          P
Constant          129.02        18.37        7.02      0.001
SqrtHt            14.5150       0.8274       17.54      0.000

S = 15.69       R-Sq = 98.4%      R-Sq(adj) = 98.1%

Analysis of Variance

Source              DF           SS          MS          F          P
Regression           1        75791       75791     307.72      0.000
Residual Error       5         1231         246
Total                6        77022
```

---

Let's focus on interpreting the coefficients and the significance tests for starters. Our estimated model could be re-written as

$$Dist = 129.02 + 14.515\sqrt{RampHt}$$

The intercept suggests that a ball rolled from a height of 0 *punti* would roll 129 *punti*, and that the distance would increase by 14.5 *punti* each time the square root of height increase by one *punto*. The significance tests strongly suggest a statistically significant relationship between the two variables and the fit is excellent ($r^2$-adj. = 98.1%).

Using the equation, we can estimate the distance of travel by simply substituting a height into the model. For instance, if the initial ramp height were 900 *punti*, we would have:

$$Dist = 129.02 + 14.515\sqrt{900}$$
$$= 129.02 + 14.515(30)$$
$$= 564.47 \ punti$$

Note that we must take care to transform our x-value here in order to compute the estimated value of y. Our result is calculated using the square root of 900, or 30.

Like any regression analysis, we must also check the validity of our assumptions about $\varepsilon$. With only seven data points in this worksheet, this example is not a good candidate for residual analysis. The next example includes that analysis, as well as another curvilinear function.

## *Another Quadratic Model*

Non-linear relationships crop up in many fields of study. Let's return to the relationship between aggregate personal savings and aggregate personal income, which we first saw in our very first linear regression session. Open the **US** worksheet.

As you may recall, these points curved around the least squares line, and there were several problems in the residuals. As a starting point, let's look at the simple linear model once more.

🖰 **Stat ➤ Regression ➤ Fitted Line Plot...** We'll use this command to run the regression, because it allows us to use quadratic and cubic transformations very easily. In this dialog, specify the two variables as shown.

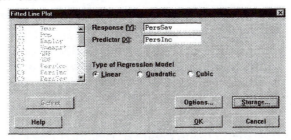

As you can see in the resulting graph, the points arc around the fitted line, and $r^2$ is only 0.58. In the Session Window, note that the estimated slope of the line is statistically significant.

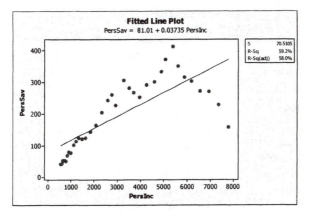

Look back at the fitted line plot. Can you visualize an inverted parabola? Such a pattern might fit this model:

$$\text{Savings}_i = \beta_0 + \beta_1\text{Income}_i + \beta_2\text{Income}_i^2 + \varepsilon_i$$

🖰 Return to the Fitted Plot command dialog again, but this time click on the option button labeled **Quadratic.** You will see another graph, get new regression output. This, then, is a short-cut way to run the regression with transformed data.

2. *How does this regression equation compare to the first one?*

3. *What do the two slopes tell you?*

4. *Are the results statistically significant?*

5. *Has the standard error decreased?*

6. *Has the coefficient of multiple determination improved?*

The simple linear model generated residuals that were approximately normal, but which formed a distinct curve when plotted versus fitted values. Let's see how the quadratic transformation affects the residuals. To generate and graph the residuals, we will use the standard regression command after we first create a new variable column containing the squared values of Personal Income.

🖰 Use the calculator to create a variable **IncSqr**, equal to **Persinc**2*.

🖰 Now run a multiple regression with **PersSav** as the response and **Persinc** and **IncSqr** as predictors, and generate the residual plots.

7. *How do these residuals compare to the earlier ones?*

8. *What are the strengths and weaknesses of the residuals?*

You may have noticed that the fitted plot dialog also features a *cubic* model option; this augments the quadratic model by adding an $X^3$ term to the equation. You might want to see if the cubic model improves the results over the quadratic.

🖰 Once more, edit the **Fitted Plot** command dialog, clicking on the radio button labeled **Cubic.**

9. *How does the cubic model compare to the quadratic?*

As you can see, the cubic model is not perfect, but this short example illustrates that data transformation can become a very useful tool. Let's turn to another example, using yet another transformation.

## *A Logarithmic Transformation*

We'll now return to the household utility dataset (**Utility**). As in prior sessions, we'll focus on the relationship between gas consumption (**GaspDay**) and mean monthly temperature (**MeanTemp**).

You may recall that there was a strong negative linear relationship between these two variables in a simple linear regression ($r^2$ was 0.85). There were some problems with the regression, though. The plot of Residuals vs. Fitted values suggested non-linearity.

When you think about it, it makes sense that the relationship can't be linear over all possible temperature values. In the sample, as the temperature rises, gas consumption falls. In the linear model, there must be some temperature at which consumption would be negative—which obviously cannot occur. A better model would be one in which consumption falls with rising temperatures, but then levels off at some point, forming a pattern similar to a natural logarithm function. The natural log of temperature serves as a helpful transformation in this case; that is, we will perform a regression with **GaspDay** as Y, and *ln*(**MeanTemp**) as X.

🖱 **Calc ▶ Calculator**. Type **LnTemp** in the box marked **Store result in**, and then type **LOGE(MeanTemp)** in the expression box. This indicates that you want to create a new variable, equal to the natural (or base e) log of temperature. Click **OK**.

🖱 Make two scatter plots that include regression lines: **GaspDay** vs. **MeanTemp** and **GaspDay** vs. **LnTemp**. The second graph is more linear than the first.

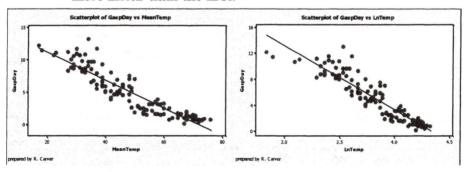

🖱 **Stat ▶ Regression ▶ Regression...** The response variable is **GaspDay**; the predictor is **LnTemp**. Generate the residual graphs.

```
Regression Analysis: GaspDay versus lnTemp

The regression equation is
GaspDay = 42.3 - 9.74 lnTemp


124 cases used, 14 cases contain missing values


Predictor      Coef   SE Coef       T       P
Constant     42.314     1.380   30.67   0.000
lnTemp       -9.7351    0.3594  -27.09   0.000


S = 1.34225   R-Sq = 85.7%   R-Sq(adj) = 85.6%
```

**10.** **What are the strengths and weaknesses of this regression?**

**11.** **What is your interpretation of the residual analysis?**

Check back through your earlier notes, and compare this regression to the simple regression involving **GaspDay** and **MeanTemp**. In our earlier model, a one-degree increase in temperature was associated with a decrease of 0.215 therms.

**12.** **What does the slope in the new model mean?**

Apply the same logic as we always have. The slope is the marginal change in $Y$, given a one-unit change in $X$. Because $X$ is the natural log of temperature, the slope means that consumption will decrease 9.735 therms when the *log* of temperature increases by 1. The key here is that one-unit differences in the natural log function are associated with ever-increasing changes in temperature as we move up the temperature scale.

You may recall from an earlier session that we once used this regression model to predict gas consumption for a month in which temperature was 40 degrees. Suppose we want to do that again with the transformed data. We can't simply substitute the value of 40, because $X$ is no longer temperature, but rather it is the natural logarithm of temperature. As such, we must substitute *ln*(40) into the estimated equation. Doing so will yield an estimate of gas consumption.

**13.** **What is the estimated consumption for a month with a mean temperature of 40 degrees?**

In the simple linear model, we obtained a negative consumption estimate for a mean temperature of 75°.

**14.** ***Estimate consumption with this new model, using the ln(75). Is this result negative?***

## Other Considerations

We are not restricted to using simple regression or to using a single transformed variable. All of the techniques and caveats of multiple regression still apply. In other words, one can build a multiple regression model that involves some transformed data and other variables as well. In addition, we can transform the response variable. However this requires additional care in the interpretation of estimates, because the fitted values must be "un-transformed" before we can work with them.

## Moving On...

### Galileo

1. Return to the data from the first Galileo experiment (first two columns), and using the Fitted Line Plot command, fit quadratic and cubic models. Discuss the comparison of the results.

2. Use the two new models to estimate the horizontal roll when a ball is dropped from 1,500 *punti.* Compare the two estimates to an estimate using the square root model. Comment on the differences among the three estimates, select the estimate you think is best, and explain why you chose it.

3. Fit a curvilinear model to the data in the third and fourth column. Use both logic and statistical criteria to select the best model you can formulate.

### Bodyfat

4. Compare and contrast the results of linear, quadratic, and logarithmic models to estimate body fat percentage using abdomen circumference as the predictor variable. Evaluate the logic of each model as well as the residuals and goodness-of-fit measures.

## Sleep

5.  Compare and contrast the results of linear, quadratic, and logarithmic models to estimate total sleep hours using gestation period as the predictor variable. Evaluate the logic of each model as well as the residuals and goodness-of-fit measures.

## Labor2

6.  Compare and contrast the results of linear, quadratic, and logarithmic models to estimate new weekly unemployment claims using the ratio of help-wanted ads to the number of unemployed persons as the predictor variable. Evaluate the logic of each model as well as the residuals and goodness-of-fit measures.

7.  Compare and contrast the results of linear, quadratic, and logarithmic models to estimate the ratio of help-wanted ads to the number of unemployed persons using the civilian unemployment rate as the predictor variable. Evaluate the logic of each model as well as the residuals and goodness-of-fit measures.

## Utility

8.  One variable in this worksheet is called **HDD** (heating degree days). It is a measure of the need for artificial heat, and equals the sum of daily mean temperature deviations below a base temperature of 65° F. Thus a month with a high value for HDD was very cold. Using **GaspDay** as the dependent variable and **HDD** as the independent, estimate and compare the linear, quadratic, cubic, and logarithmic models. Which model seems to be best?

## Bowling

These are results from a bowling league. Each person bowls a "series" consisting of three "strings" (maximum score = 300 per string).

9.  Suppose we want to know if we can predict a bowler's series total score based on the score of his or her first string. Compare a linear model to another model of your choice, referring to all relevant statistics and graphs.

## Output

10. Construct linear, cubic, and logarithmic models with the Index of industrial production as the dependent and durables production as the independent variable. Compare the strengths and weaknesses of the models.

11. Construct linear, cubic, and logarithmic models with durables production as the dependent and nondurables production as the independent variable. Compare the strengths and weaknesses of the models.

12. Construct linear, cubic, and logarithmic models with the consumer goods production as the dependent and durables production as the independent variable. Compare the strengths and weaknesses of the models.

264

# Basic Forecasting Techniques

### *Objectives*

In this session, you will learn to:
- Identify common patterns of variation in a time series
- Use moving averages to make and evaluate a forecast
- Use exponential smoothing to make and evaluate a forecast
- Using trend analysis to make and evaluate a forecast

### *Detecting Patterns over Time*

In the most recent sessions, we have been with building models that attempt to account for variation in a dependent, or response, variable. These models, in turn, can be used to estimate or predict future or unobserved values of the dependent variable.

In many instances, though, variables behave fairly predictably over time. In such cases, we can use techniques of *time series forecasting* to predict what will happen next. There are many such techniques available; in this session, we will experiment with only three of them. This session is very much an introduction to these tools; consult your textbook for additional techniques.

Recall that a *time series* is a sample of repeated measurements of a single variable, observed at regular intervals over a period of time. The intervals could be hourly, daily, monthly; what is important is that it be regular. When we examine a time series, we typically expect to find one or more of these common idealized patterns, often in combination with one another:

- ***Trend***: a general upward or downward pattern over a long period of time, typically years. A time series showing no trend is sometimes called a *stationary time series.*
- ***Cyclical*** *variation*: a regular pattern of up-and-down waves, such that peaks and valleys occur at regular intervals. Cycles emerge over a long period of years.
- ***Seasonal*** *variation*: a pattern of ups and downs within a year, repeated in subsequent years. Most industries have some seasonal variation in sales, for example.
- ***Random, or irregular***, *variation*: movements in the data, which cannot be classified as one of the preceding types of variation.

Let's begin with some real-world examples of these patterns. Be aware that the four time series components just listed are ideal categories. It is rare to find a real time series which is a "pure" case of just one component or another.

## Some Illustrative Examples

Open the file **US**. All of the variables in this file are summarized or aggregated annually. Therefore we cannot find seasonal variation here. However there are some good examples of trend and cycle, as well as irregular variation.

**Graph ➤ Time Series Plot...**  Choose a simple plot. In the dialog box we'll select the variables **NHMort, Unemprt, Starts, M1,** and **Pop**. This will create five graphs.

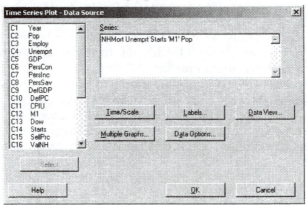

The five variables are these (1965–1999):
- **NHMort**: New Home Mortgage Interest Rate
- **Unemprt**: Unemployment Rate (%)
- **Starts**: Housing Starts (thousands), i.e., the number of new homes on which construction started in the year
- **M1**: Money Supply (billions of $)
- **Pop**: Population of the US (thousands).

The topmost graph shows the population of the United States, and one would be hard-pressed to find a better illustration of a linear trend. During the period of the time series, population has grown by a nearly constant number of people each year. It is easy to see how we might extrapolate this trend.

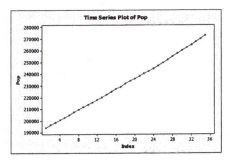

In the next graph (M1), there is also a general trend here, but it is *non*-linear. If you completed the session about non-linear models, you might have some ideas about a functional form that could describe this curve. In fact, as we'll see later, this graph is a typical example of growth that occurs at a constant *percentage* rate, or exponential growth.

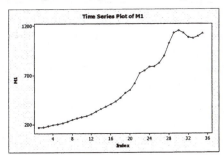

The third graph in the stack (housing starts) is a rough illustration of *cyclical* variation, combined with a moderate negative trend. Although the number of starts increases and decreases, the general pattern is downward, with peaks and valleys spaced fairly evenly.

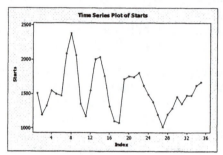

The fourth graph, showing the unemployment rate during the period, has unevenly spaced peaks, and the upward trend visible on the left side of the graph appears to be declining on the right side. The irregularities here suggest a sizable erratic component. This graph also illustrates the way various patterns might combine in a graph.

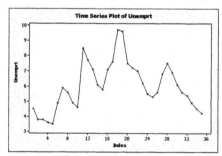

Finally, the graph of mortgage rates is almost entirely irregular movement. The pattern that we see is not easily classified as one of the principal components noted earlier.

To see seasonal variation we return to the **Utility** file, with home heating data from New England.

🖰 Open the worksheet **Utility**. We'll focus on the variable **MeanTemp**, which represents the mean temperature for the month. It certainly makes sense to expect that we'll see seasonal variation.

🖰 **Graph ➤ Time Series Plot...** We'll make another simple plot, but this time, to help us see the seasonal periodicity, we'll add some lines to the graph. Select the variable **MeanTemp** and then click on the word **Time/Scale** button in the dialog box. Select the **Reference lines** tab. This brings up the dialog shown below.

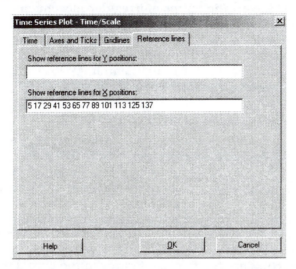

We want to insert a vertical line at each January observation in the graph. It happens that January, 1991 was the fifth observation, so we'll specify that we want a line at the fifth point, the seventeenth point, and then every twelfth observation thereafter.

    As shown below, enter these values in the box marked **Show reference lines for X positions**. Be sure to leave a space between each value:

5   17  29  41  53  65  77  89  101  113  125  137

    Click **OK**. Click **OK** in the main Time Series Plot dialog as well.

Look at the graph. Note that the first "valley" occurs at the fifth observation (January, 1991). Now scan across to twelve months later, at observation 17. Do the same all the way across the graph. This is a classic illustration of regular seasonal variation.

We can exploit patterns such as these to make predictions about what will happen in future observations. Any such attempt to predict the future will of course yield imperfect forecasts. Because we cannot eliminate errors in our predictions, the trick in forecasting is to *minimize* error. Each of the techniques illustrated below is suited to modeling different time series patterns, and each has its virtues. Each also inevitably involves *forecast errors,* which are the differences between a forecast value and the (eventual) observed value.

The first two techniques we'll explore are most useful for "smoothing out" the erratic jumps of irregular movements. They are known as *moving averages* and *exponential smoothing.* For each of the techniques that follows, we'll illustrate the technique using a single time series, the dollar value (in millions) of exports of domestic agricultural products from January, 1948 through March, 1996. That time series is found in the file called **Eximport**. Open that file.

## Forecasting Using Moving Averages

Moving averages[1] is a technique that is appealing largely because of its simplicity. We generate a forecast merely by finding the simple average of a few recent observations. The key analytical question is to determine an appropriate number of recent values to average. In general, we make that determination by trial and error, seeking an averaging period that would have minimized forecast errors in the past.

Thus, in a Moving Average analysis, we select a moving average 'length' or interval, compute retrospective "forecasts" for an existing set of data, and then compare the forecast figures to the actual figures. We summarize the forecast errors using some standard statistics explained below. We then repeat the process several more times with different moving average intervals to determine which interval length would have performed most accurately in the past.

Before making any forecasts at all, let's look at the time series.

🖰 Construct a time-series plot of **EXag**.

1. ***Comment on any patterns or unusual features of this plot.***

🖰 **Stat ▶ Time Series ▶ Moving Average...** In the dialog, select the variable **EXag.** Specify a **MA Length** of 6 months, and indicate that you wish to generate 1 forecast (for April, 1996). Also click on the **Time...** button, choose **Stamp**, and select the **Date** variable in the **Stamp** box.

---

[1] In the remainder of the session we will assume that your primary text covers the theory and formulas involved in these techniques. Consult your text for questions about these issues.

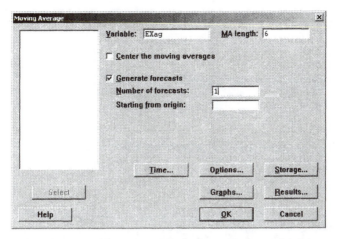

Maximize the resulting graph window. Due to the large sample the graph is crowded, but you can distinguish between the observed values (black) and the smoothed six-month moving average series (red). Note that the red line 'shadows' the black, slightly to the right. Note also that the red line is smoother than the black, never quite reaching the extreme values of the black dots. At the far right is our single forecast, shown as a tiny green diamond bracketed by two blue triangles. Checking the Session Window, we find that the estimate for April 1996 is for $4,851 million in agricultural exports.

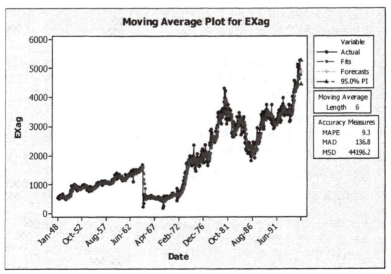

```
Moving Average for EXag

Data        EXag
Length      579
NMissing    0

Moving Average
Length   6

Accuracy Measures
MAPE        9.3
MAD       136.8
MSD     44196.2

Forecasts
Period  Forecast    Lower     Upper
580          4851  4438.96  5263.04
```

In the Session Window and on the graph, we find the three summary statistics characterizing the historical accuracy of this model. These measures are:

- **MAPE**: Mean Absolute Percentage Error, representing the mean forecast error, expressed as a percentage of the actual observed values. This number is fairly easy to interpret: In this case, on average, our forecast values were "off" by about 9.3% from the actual values.
- **MAD:** Mean Absolute Deviation, simply representing the mean of the absolute values of the forecast errors, in this case $136 million.
- **MSD:** Mean Squared Deviation, which is the sum of the squared forecast errors, divided by *n*. This is very similar to MSE, excepts that MSE divides by the number of forecasts.

In general, we are looking for a *moving average length* that minimizes these measures of forecast error. Let's generate two more analyses and select the one with the best error statistics.

🖱 Edit the moving average dialog and specify a 3-month average period.

🖱 Edit the dialog again, this time for a 12-month analysis.

🖱 Print all three graphs.

2.   *Which of the three averaging periods is best? What is the one-month forecast using the best model?*

## *Forecasting Using Exponential Smoothing*

One drawback of moving averages is that, ultimately, our forecast is based on a very few numbers. We effectively discard most of the sample as we generate a single forecast. Exponential smoothing manages to incorporate the entire dataset into each forecast. This technique smoothes out a jagged time series by creating forecasts that represent a weighted average of the prior forecast and the prior observation. The general model is as follows:

$$F_{t+1} = \alpha Y_t + (1-\alpha)F_t$$

where   $F_t$ is the forecast value for time period $t$
$Y_t$ is the actual series value for period $t$
$\alpha$ is the *smoothing constant* such that $0 < \alpha < 1$

Our job is to select a smoothing constant that provides the minimum forecast error.

🖰 **Stat ➤ Time Series ➤ Single Exp Smoothing...** As shown, select the new variable, **EXag**, and specify a 'smoothing constant' of 0.5. Generate one forecast. As above, click **Time...** and stamp the horizontal axis with the **Date**.

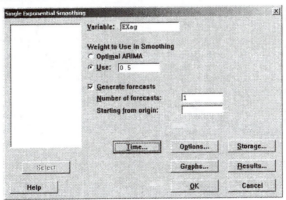

The output is very similar to that of the Moving Average analysis. In terms of MAPE, this choice of a smoothing weight looks relatively good. We could try several more, or we can request that the software optimize the choice of weight for us. Let's do that.

🖱 Edit the last command dialog, and click the option button marked **Optimal ARIMA,** and click OK.

> **3.   *This graph is the best we can do with Exponential Smoothing. How do these results compare to the moving average that you chose?***
>
> **4.   *How do the one-month forecasts compare to each other?***

Unfortunately, the exponential smoothing command requires a complete set of data, so if a time series contains any missing observations, we cannot use the command.

## Forecasting Using Trend Analysis

It is clear from the graphs that there has been a general upward trend in the dollar value of agricultural exports since the late 1940's. Let's try to model that trend in two ways.

🖱 **Stat ▶ Time Series ▶ Trend Analysis...** Select **EXag** again, specify a **Linear** trend, and ask for one forecast value.

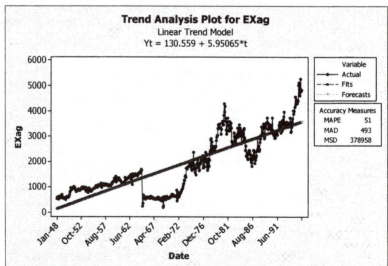

From this graph we immediately see that the linear trend model is a poor choice. The summary statistics are weak, the actual series oscillates around the trend line, and the forecast value appears to be very unlikely. Note however how clearly you can see the cyclical pattern in the data.

If a linear model is a poor fit, let's try some non-linear trends.

⌕ Edit the trend analysis dialog twice, first choosing the **Quadratic** model, and then the **Exponential Growth** model.

5. *Which of the resulting analyses is better?*

6. *Why did you select that one?*

## Moving On...

Using the techniques presented in this session, answer the following questions. Whenever you present a conclusion, be sure to explain what technique you used, and how you arrived at your conclusion.

### Output

These questions focus on **ProdInd**, which is an *index* of total industrial production in the US, much like the Consumer Price Index is a measure of inflation.

1. Which of the four components of a time series can you see in this series? Explain.

2. Generate 6, 9, and 12-month moving averages for this series. Which averaging length would you recommend for forecasting purposes, and why? Generate one forecast using that averaging length.

3. Compute one optimal Exponential Smoothing forecast for this variable.

4. Select a trend model that you feel is appropriate for this variable, and generate a forecast.

5. Compare your various forecasts. Which of the predictions would you rely upon, and why?

### Utility

6. Generate a three-month moving average forecast for the variable **KWHpDay**, which is the mean daily amount of electricity usage in my home. Given the pattern in the entire graph, explain why it may be unwise to rely on this forecast.

7.  Why might it be unwise to use a few recent months of data to predict next month's usage? What might be a better approach?

## US

8.  Perform a trend analysis on the variable **M1**, using linear, quadratic, exponential, and S-curve models. Which model seems to fit the data best? Why might that be so?

9.  Among the columns in this dataset find another variable that is well modeled by the same function as **M1**. Why does this variable have a shape similar to **M1**?

## Eagles

This worksheet contains the number of breeding pairs of bald eagles in the contiguous United States between 1963 and 1998.

10. Create a time series plot of the number of breeding pairs, and comment on what you see.

11. Generate a 4-year moving average for this series; comment on noteworthy features of your results. Would you use this forecasting method to predict future years?

12. Use the trend analysis command to fit a reliable predictive model to this set of data. Which model do you find fits the data best? Use your model to estimate the number of breeding pairs for 1980, a year in which no figure was available.

## Labor1

This worksheet contains monthly labor force participation data for the United States from 1948 to 1996.

13. Columns 4 through 6 contain the percentage of eligible males, females, and teens who participate in the labor force each month. Create three time series plots, and describe the story that each graph tells.

14. Use one or more of the forecasting techniques presented in this session to make a one-period forecast for teenage labor force participation. Say which method you have selected and explain why.

15. The last variable in the worksheet represents the average number of work hours per week in manufacturing jobs. Which common time series patterns do you see in this particular series?

16. Create an optimal exponential smoothing forecast for one period for the **HoursM** variable, and explain your results.

## Terrorism

Extracted from a 2001 CIA report on international terrorism, this worksheet contains the CIA's estimates of the number of terrorist incidents occurring between 1971 and 1996.

17. Fit an appropriate trend model to this set of data, and explain which model you have selected and why.

18. Use your model to estimate the number of international terror incidents for 1997.

19. Clearly, terrorist activity increased dramatically in the years following the CIA report. What does this example illustrate about the risks of using time-series forecasts in the absence of an understanding of underlying causes of a phenomenon?

# Nonparametric Tests

## Objectives:

In this session, you will learn to:
- Perform a sign test for a median
- Perform a Wilcoxon signed rank test for a median
- Perform a Mann-Whitney U test comparing two medians
- Perform a Kruskal-Wallis test comparing two or more medians
- Compute Spearman's rank order correlation coefficient
- Perform a runs test for randomness

## Nonparametric Methods

Many of the previous sessions have illustrated statistical tests involving population parameters such as $\mu$, and which often require the assumption that the underlying population is normally distributed. Sometimes we cannot assume normality, and sometimes the data we have do not lend themselves to computing a mean (for example, when the data are ordinal).[1] The techniques featured in this session are known as *nonparametric methods*, and are applicable in just such circumstances. In particular, we will mostly use them when we cannot assume that our sample is drawn from a normal population.

Generally, it is preferable to use a parametric test when the population is close to normal, because such tests tend to be more discriminating and powerful. Using a nonparametric test inappropriately increases the risk of a Type II error. Despite that risk, nonparametric tests are easy to use and interpret, and free us from the small-sample

---

[1] Consult your text to define or review the *scales of measurement* of data.

restrictions on non-normal population data. When our data do not satisfy the basic assumptions that underlie the tests we have already seen, we should select a nonparametric alternative. Although there are a large number of nonparametric methods, this session will illustrate a few of the more common elementary techniques.

## A Sign Test

Perhaps the simplest of these techniques is the sign test. We use it to test hypotheses concerning the median of a population. The test relies on the idea that about half of the observations in a random sample should lie above the population median. Thus, with respect to a hypothetical value of the median, the sample is a binomial experiment of $n$ trials, each with a 0.5 probability of success, and each independent.

An example will illustrate. You may recall the dataset containing the cholesterol readings for heart attack patients. In addition to the heart attack patients, there was a healthy control group. We'll begin by testing whether the median cholesterol for the control group was less than 200. Following the logic just outlined, if 200 is indeed the population median, we expect about half of the sample observations to be above 200. Formally, we'll test:

$H_o$: Median $\geq$ 200 vs.
$H_A$: Median < 200

This is similar to a test we performed in Session 10 but there are two important differences. That test dealt with the population *mean* rather than the median, and if we had a small sample we needed to assume the population was normally distributed.

⌐🖰 **File ➤ Open Worksheet...** Open the file **Cholest**.

⌐🖰 **Stat ➤ Basic Statistics ➤ Graphical Summary...** Select the **Control** variable. This will shed light on the assumption of a normal population.

In the resulting output (on the next page), we see that there is some question about whether the population is normally distributed. With a sample of just 30 the $t$ test might be inappropriate. Therefore, this is a candidate for a sign test.

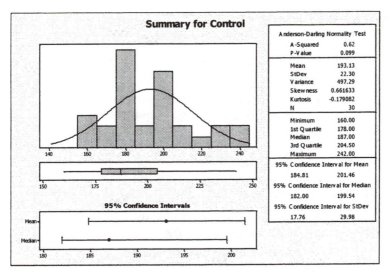

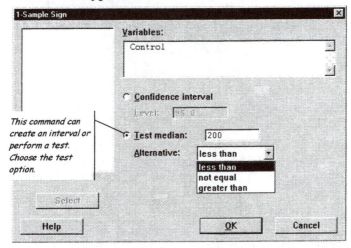

🖰 **Stat ➤ Nonparametrics ➤ 1-Sample Sign...** In the dialog, select **Control**, specify a median of 200, and select a **less than** alternative hypothesis.

The Session output on the next page shows that 20 observations were below 200, 2 were equal to 200, and only 8 observations were above it. If the null hypothesis were true, this would happen by chance less than 2% of the time ($P \approx 0.0178$). This constitutes strong evidence that the population median is less than 200.

---

**Sign Test for Median: Control**

Sign test of median =  200.0 versus < 200.0

|          | N  | Below | Equal | Above | P      | Median |
|----------|----|-------|-------|-------|--------|--------|
| Control  | 30 | 20    | 2     | 8     | 0.0178 | 187.0  |

---

We can also use the sign test to do the equivalent of a paired-sample *t* test test. In this case, we have repeated observations of cholesterol readings for the heart attack patients. Suppose we want to know if their readings declined in the first few days after their attacks. We might want to test whether the median difference between the Day 2 and Day 4 readings is negative. First, we must compute the differences in readings for each patient.

🖰 **Calc ➤ Calculator** Specify a new variable (call it **Change**), which equals **4-Day – 2-Day** and click **OK**.

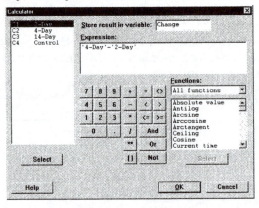

🖰 **Stat ➤ Nonparametrics ➤ 1-Sample Sign...** In the dialog, select **Change**, specify a median of **0**, and again select a **less than** alternative. This will test whether the median change was negative, i.e., whether more than half of the patients experienced a decline in cholesterol.

---

**Sign Test for Median: Change**

Sign test of median =  0.00000 versus < 0.00000

|         | N  | Below | Equal | Above | P      | Median |
|---------|----|-------|-------|-------|--------|--------|
| Change  | 28 | 19    | 0     | 9     | 0.0436 | -19.00 |

---

Here, the results are consistent with the alternative hypothesis: about twice as many patients (19 compared to 9) had decreased cholesterol levels. The *P*-value of 0.0436 indicates that we reject the null hypothesis at the 0.05 significance level.

## A Wilcoxon Signed Rank Test

The sign test compares observations based on their *signs*: Does the observation lie above or below the median? As far as the sign test is concerned the observation closest to, but above, the median is equivalent to the observation farthest above. The Wilcoxon Signed Rank test (also known as the *signed rank test*) takes the magnitudes of the deviations into account by ranking them. This test requires a reasonably symmetric (though not necessarily normal) population.

In this test, the idea is that when we sample from a symmetric population, we should find that the observations are dispersed symmetrically about the population median. This test looks for that symmetry by comparing the rankings of sample observations on each side of the median. It does so by computing all of the individual deviations from the median, ranking them from smallest to largest in absolute value, and dividing them into positive (above the median) and negative. Then it sums the positive and negative ranks, and compares the two sums. If the population is symmetric and the hypothesized median is correct, the two sums should be very close. In practice, it's simpler than it sounds. Let's apply this test to our first set of data.

🖰 **Stat ➤ Nonparametrics ➤ 1-sample Wilcoxon…** Just as with the Sign test, select the variable **Control**, and test for a **Median** equal to **200** against a **less than** alternative. The results are shown here:

**Wilcoxon Signed Rank Test: Control**

```
Test of median = 200.0 versus median < 200.0
```

|         | N  | N for Test | Wilcoxon Statistic | P     | Estimated Median |
|---------|----|-----------|--------------------|-------|------------------|
| Control | 30 | 28        | 127.5              | 0.044 | 191.0            |

In this test, two of the thirty observations are omitted because they are equal to the hypothesized median value of 200. Based on the remaining 28, the procedure computes the test statistic of 127.5, which is significant at the 0.05 level (*P* = .044).[2] Thus taking ranks into account, we reject the null hypothesis.

---

[2] Consult your text or Minitab help for the computational details.

We can also use the Wilcoxon signed rank test as an equivalent to the paired $t$ test, just as we did the sign test. As an exercise, perform a Wilcoxon signed rank test to evaluate the null hypothesis that cholesterol levels remain stable or increase from Day 2 to Day 4 among heart attack patients.

1.   ***Compare the results of this Wilcoxon test to those of the earlier sign test. Comment on noteworthy similarities and differences between the results of the two tests.***

## Mann-Whitney U Test

The Mann-Whitney U Test is the nonparametric version of the independent samples $t$ test, but once again is a test of medians. In particular, we use this test when we have two independent samples and can assume that they are drawn from populations with the same (but not necessarily normal) shape and dispersion.

Like the previous tests the U Test can be used for ordinal, interval, or ratio data, and is based on rankings. The underlying strategy is equivalent to the Wilcoxon signed rank test. If the medians of the two populations are equal, and the shape of the two populations is the same, then when we pool and rank all of the observations, the rankings should be balanced across the two samples.

Consider the data in the worksheet called **StateTrans**. Among the variables in the file are the number of fatal auto accidents (**FatCrash**) and a categorical variable (**AVMCat**) indicating whether the state ranks in the upper or lower half of the U.S. in vehicle miles traveled per capita. In other words, we can divide the states into those whose citizens drive a lot, and those who do not. In this test, we will hypothesize that the high-traffic states will have more fatal crashes than the low-traffic states.

Before performing the test, we want to check the assumptions of similar shape and variance. We can do so by consulting the descriptive statistics. Open the worksheet called **StateTrans**.

   ⬦ **Stat ➤ Basic Statistics ➤ Graphical Summary ...** Select the variable **FatCrash**, and request graphs by **AVMCat**.

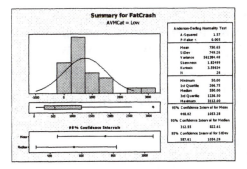
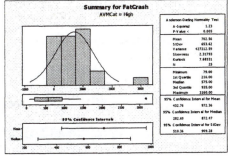

Here we see two distributions of comparable shape, but *not* normal. Their variances are similar, and so meet the requirements of the U test more nearly than those of the 2-sample t test.

To perform a Mann-Whitney test, our samples must be in separate columns of the worksheet; that is they must be "unstacked."[3] In our worksheet, the number of fatal accidents and the vehicle-miles category are represented as two variables. For this test, we need to think of the annual vehicle-miles traveled as coming from two separate populations, and create two new columns of accident data, representing the low- and high-travel states. We do so as follows:

🖰 **Data ➤ Unstack Columns...** As shown below, specify that you want to **Unstack the data in FatCrash**, separating it into two new columns using the subscript data in **AVMCat**.

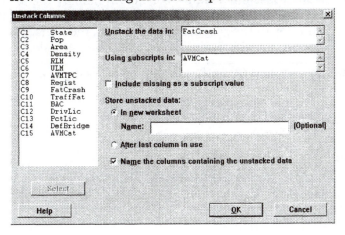

---

[3] For more information about "unstacked" data, see Appendix C.

This creates a new worksheet with two columns, separating the accident data for the low- and high-travel states. Notice that Minitab automatically names the new variables appropriately. Now we can perform the test.

🖰 **Stat ➤ Nonparametrics ➤ Mann-Whitney…** The first sample is **FatCrash_High**, and the second is **FatCrash_Low**. Choose a **greater than** alternative hypothesis.

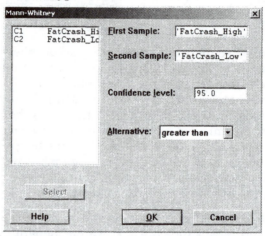

The results (Session Window; see below) display the sample medians for the two groups, and the difference between them (-0.15). ETA1 and ETA2 refer to the respective population medians. The test statistic W is the sum of the rankings from the first sample.

```
Mann-Whitney Test and CI: FatCrash_High, FatCrash_Low

                    N   Median
FatCrash_High      25    575.0
FatCrash_Low       26    550.0

Point estimate for ETA1-ETA2 is 19.5
95.1 Percent CI for ETA1-ETA2 is (-283.9,265.9)
W = 657.0
Test of ETA1 = ETA2 vs ETA1 > ETA2 is significant at 0.4513
```

Minitab computes a confidence interval for the difference in medians and reports the P-value for the test.[4] In this example, the test is significant at the 0.4513 level, which is to say that we do not reject our null hypothesis of equal variances at the customary 0.05 level. The sample data indicate no statistically significant difference between the median number of fatal accidents in high- and low-travel states.

## Kruskal-Wallis Test

Whereas the Mann-Whitney test addresses the comparison of central location for two non-normal populations, the next test does the same for *two or more* non-normal populations. Thus, the Kruskal Wallis test is the nonparametric analogue of the one-factor ANOVA.

The Kruskal Wallis H-test assumes that the samples are drawn from independent populations with identical shape and spread. Unlike the one-way ANOVA, there is no need to assume normality. The strategy is once again based on a comparison of pooled rankings, and the null hypothesis is that the medians of the $k$ populations are equal; the alternative hypothesis is that at least one of the medians is different.

To illustrate, let's consider the data from Dr. Stanley Milgram's famous experiments on obedience to authority, conducted in the early 1960's at Yale University. Under various conditions, subjects were instructed to administer electrical shocks to another person. In reality, there were no electric shocks, but subjects believed that there were (see Session 14 for a more detailed description of the experiments). One factor in these experiments was the proximity between the subject and the person "receiving" the shocks. We'll find out if proximity to the victim had a significant impact on the maximum voltage of the shock delivered. As before, we'll check our assumption of normality before conducting the statistical test. Open the **Milgram** worksheet.

2. *Create graphical summaries for the variable* Volts, *grouping by the variable* Proximity. *Do any or all of these four distributions appear to be normal? Explain.*

🖱 **Stat ➤ Nonparametrics ➤ Kruskal-Wallis...** The **Response** variable is **Volts**, and the **Factor** is **Proximity**.

---

[4] With other datasets you might see two reported P-values. The first P-value is based on all of the sample data; the second adjusts the P-value in the event that there were ties in the rankings between the two samples. The unadjusted value is conservative (i.e., higher) when there are ties; the adjusted P-value is, however, the more accurate of the two when there are ties.

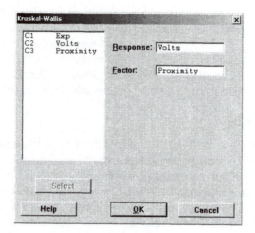

Once again, the results are in the Session Window:

**Kruskal-Wallis Test: Volts versus Proximity**

```
Kruskak-Wallis Test on Volts

Proximity       N  Median  Ave Rank      Z
Remote          40   450.0    101.7    3.34
Voice           40   450.0     91.3    1.70
Same Room       40   307.5     71.2   -1.47
Touch Hands     40   195.0     57.9   -3.56
Overall        160             80.5

H = 21.68  DF = 3  P = 0.000
H = 24.87  DF = 3  P = 0.000  (adjusted for ties)
```

In the output we see the median voltage administered for each experimental proximity condition as well as the mean ranking for each. The test statistic, which follows a Chi-square distribution with $k-1$ degrees of freedom, in this case is $H = 21.68$ which has a $P$-value approximately equal to 0.000. As in the Mann-Whitney test, the test statistic is also adjusted in the event of tied rankings. The conclusion in this particular example is that the proximity between the subject and the "victim" had a significant effect on the maximum amount of shock the subject was willing to administer.

### *Spearman Rank Order Correlation*

In one of the earliest sessions, we learned to compute the Pearson correlation coefficient, which measures the linear association between two interval or ratio variables. Sometimes, we have only ordinal data but may still suspect a linear relationship. For example, we may want to compare the order of finish for runners in two qualifying races.

The Spearman rank-order correlation coefficient is simply the familiar Pearson *r*, applied to ordinal data. For example, we have some basic geographic and population data about the state of Texas (**TexasGeog**). In that dataset, we have the population figures for each county in the state as recorded by the 1990 and 2000 U.S. Census. Suppose we wanted to determine the extent to which the rankings of the counties changed, if at all, in the decade. We'll rank the counties for each of the years, and then compute the correlation coefficient for the ranks.

🖱 **Data ➤ Rank** As shown in the dialog, we want to rank the counties by **Pop1990**, and place the rankings into a new column that we'll label **Rank90**. Each county will be assigned a rank, with 1 assigned to the smallest county (Loving County).

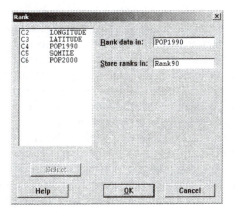

🖱 Now rank the states by **Pop2000**, creating a new column called **Rank2000**.

A brief look at the two new columns in the Data Window suggests that there were slight changes in the rankings. Let's see how close the two lists are:

🖱 **Stat ➤ Basic Statistics ➤ Correlation...** This is a familiar dialog. Select the two rank variables, and click **OK**.

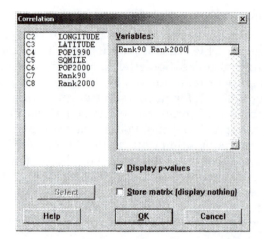

In the Session Window, you'll find that the correlation between the ranks is 0.925, which is significantly different from zero.

## A Runs Test

The final test in this session is a test for randomness. Often, we may want to ask whether observations in a sequence are independent of one another. For instance, we may have reason to think that an observation at time $t$ was correlated with the observation at time $t$–1. The Runs test checks a series for randomness by counting the number of "runs" in the data. A run is defined as a consecutive group of observations sharing a particular characteristic of interest, like consecutive heads or tails when flipping coins.

We know from studying binomial distributions that we can reasonably expect occasional runs in any Bernoulli process with independent trials. However, we also know that there are probability distributions pertaining to such patterns, and that some run patterns would stretch credulity that the data were randomly generated. The runs test is based on that fact.

Let's consider two examples. The first will use simulated data, and we'll use the test to see if indeed Minitab's random data generator creates a random sequence.

 **Calc ➤ Random Data ➤ Bernoulli...** Generate 100 rows of data, with a probability of success = .5. This will create a randomly generated column of 0s and 1s. Store the results in **C10**.

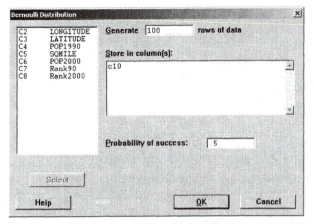

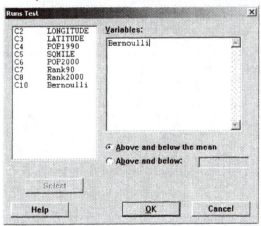

   In the data window, label **C10** as **Bernoulli**.

   **Stat ➤ Nonparametrics ➤ Runs Test...** Select **Bernoulli**.

The runs test output appears in the Session Window; naturally, your random data and test results will differ from those shown in the illustration on the next page. The test compares the expected number of runs in a series of 100 observations to the observed number, and computes a significance level. In this example, the value K = 0.48 is the sample mean, and 48 observations lay above it. We expected 50.92 runs and observed 58. The null hypothesis of a random arrangement cannot be rejected, which is to say that we persist in our assumption of a random sequence.

```
Runs Test: Bernoulli
Runs test for Bernoulli

Runs above and below K = 0.48

The observed number of runs = 58
The expected number of runs = 50.92
48 observations above K, 52 below
P-value = 0.154
```

### 3.   *What did your simulation show?*

The second example will use a real time series, to see if a particularly "noisy" series follows a random pattern. One word of warning when applying the Runs Test in Minitab: Your variable may have missing cases at the beginning or end of the series, but *not* anywhere in the middle. In this example, the variable is the percentage of industrial capacity utilized each month in the United States over a period of years. Open **Output**. The variable of interest is called **CapUtil**.

The time series itself looks like this:

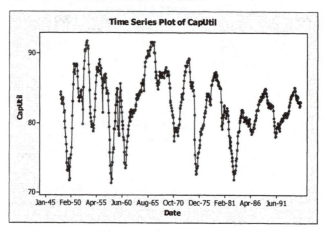

The series is quite erratic, and we might hypothesize that it is a random series. The Runs test can evaluate that hypothesis.

🖱 **Stat ➤ Nonparametrics ➤ Runs Test...** Select **CapUtil**.

### 4.   *Based on this test, how many of the observations were above and below the sample mean? Does this test conclude that the time series is or is not random?*

## Moving On...

Apply the techniques learned in this session to the questions below. In each case, indicate which test you have used, and cite appropriate test statistics and p-values.

### Output

1. Is the capacity utilization among manufacturing firms (**CapUtMfg**) a random series? Explain your reasoning.

2. Use two different approaches to test the hypothesis that the median manufacturing capacity utilization is 80%. Would a $t$ test be more or less appropriate than a nonparametric test in this case? Explain.

### Swimmer1

3. Did swimmers tend to improve between their first and last recorded time, regardless of event?

4. Looking only at second race data for the 100-meter events, are the median times the same regardless of event?

### Swimmer2

5. Did swimmers tend to improve between their first and second heats in the 50-meter freestyle?

6. Looking only at second race data, do those who compete in both events swim faster in the 50-meter freestyle than they do in the 50-meter backstroke?

### StateTrans

7. Is there a significant rank-order correlation between the urban lane miles in a state and the density of the state?

8. All of the U.S. states use blood alcohol measurements as part of the legal definition of driving while intoxicated. All of the states use of one of two standards. Do states with a stricter blood alcohol limit (lower BAC value) have a significantly lower number of automobile fatalities than those who use the other limit?

## Violations

9. One might expect that through four years of college, students would learn ways to avoid parking citations on campus. Is there a significant difference in the median fines paid by Freshman, Sophomore, Junior, and Senior students? Explain your findings.

## LondonNO

This file contains hourly readings of Nitric Oxide, measured in West London. Each column contains daily readings at a given hour of the day.

10. Based on older observations, the median concentration of NO at 7:00 AM is approximately 19 parts per billion. Is there significant evidence in this dataset that the median concentration at 9:00 AM exceeds 19 ppb? Explain.

11. Does the NO concentration decrease between 9:00 AM and 11:00 AM?

## Haircut

12. Does the price a student pays for a haircut vary by gender?

13. Is home region (urban, suburban, or rural) a statistically significant factor that affects the price students pay for haircuts?

# Session 22

## Introduction to Design of Experiments

### *Objectives*

In this session, you will learn to:

- Understand the goals of designed experiments
- Create designs for one-, two- and multi-factor experiments
- Understand the function of replication and blocking
- Create fractional designs for multi-factor experiments
- Create Plackett-Burman designs

### *Experimental and Observational Studies*

Recall the first example of Session 1. We analyzed the results of a simple classroom experiment with paper helicopters to compare flight times of helicopters with and without attached paperclips. We have revisited the experimental data a few times, but have not devoted very much attention to the concept and design of the experiment itself.

Most of the datasets accompanying this book are *observational* in nature—they record counts and measurements from populations or processes as we find them. In most instances, no researcher intervened except to record the data. As we have seen, we can perform descriptive and inferential analysis with observed data.

In many cases, though, it is to our advantage to *experiment*: to manipulate *factors* intentionally so that we can assess and understand their impact on a particular *response variable*. In the case of the helicopters, for instance, we systematically flew the helicopters with and without the paperclips (factor) so that we could measure the differences in flight duration (response variable).

The previous sessions have dealt solely with data analysis. In this session, we'll see how Minitab can help us to design experiments that generate data for later analysis. You should be able to recognize the appropriate analytical approach for most of the designs that we'll investigate in this session. The emphasis here is on the choice and creation of appropriate designs, rather than on the analysis of the resulting data.

## Goals of Experiments

Montgomery (1991, p. 8) defines the *statistical design of experiments* as "the process of planning the experiment so that appropriate data that can be analyzed by statistical methods will be collected, resulting in valid and objective conclusions." In other words, one major objective in the statistical design of experiments is to ensure that our experiments will yield useful data that can be analyzed to allow us to draw valid conclusions and make better decisions.

Hoerl and Snee (2002), referring to industrial experimentation, identify three experimental environments, suggesting three broad goals for experimentation. These categories are relevant to experiments in the natural and social sciences as well. Each environment is characterized by its own fundamental question, though any particular experiment might be a hybrid of the three categories.

- *Screening:* Which variables have the largest effect on the response variable?
- *Characterization:* How large are the effects of the few most important variables? Are there any important interactions among these variables? Are relationships curvilinear?
- *Optimization:* At what values of the independent and control variables does the response variable attain a desired minimum or maximum?

Consider the paper helicopters again. Suppose that our ultimate goal is optimization: to design a helicopter that remains airborne longer than any other when dropped from a given height. How do we do this? There are many factors that might influence duration: type of paper, wing length, weight, body width, the proportion of length to width, and others. A *screening* experiment would seek to identify a few key factors affecting the duration of a flight. A *characterization* experiment would focus on factors identified by screening, and seek to understand how they collectively affect flight duration. With the relevant variables identified, we would finally use an *optimization* approach to fine-tune the variables to design a long-flying helicopter. We'll begin with some standard approaches to screening experiments, and we'll continue with our classroom paper helicopter example.

## Factors, Blocks, and Randomization

We'll refer to the helicopter features that might affect flight time as *factors*. In general, a factor is a variable that can affect the response variable of interest. Factors may be qualitative or quantitative.

It is evident that we can control some of the factors that affect flight duration, but we cannot control others. Surely, we can establish a standard height from which to launch the helicopters and we can select the type of paper to use and so forth. We can run our tests indoors to avoid the uncertainties of wind and weather. However, individual students may drop their helicopters in slightly different ways, or perhaps one student is more accurate with a stopwatch than another. These individual differences can influence the results of our experiments. When we design an experiment, we ideally take into account *all* potential sources of variation in the response variable. We generally do this in three ways.

- We deliberately manipulate factors (sometimes called *treatments*) that we believe affect the response variable. We may repeat, or *replicate*, measurements of the response variable for a given treatment condition through a series of experimental runs. [1] We do this to gauge and control for measurement error.

- We *block* some runs to account for expected uncontrollable similarities among runs. We block when we have good reason to expect that some measurements will tend to be similar due to the factor being blocked. For example, we might group all helicopter times that were recorded by the same time-keeper.

- We *randomize* experimental runs to account for those factors that we cannot block or manipulate. Thus, if we expect that differences among individual students might affect flight times, we randomly assign students specific helicopter designs. Similarly, if each student were going to test several designs, we would randomize the sequence of flights, so that each student would drop the helicopters in a different order.

---

[1] A *run* in this context is one repetition of one experimental condition. In some other contexts (e.g. nonparametric methods) the word run might refer to a sequence of similar values, like a series of positive residuals.

## *Factorial Designs*

One might be inclined to conduct a series of experiments, each focusing exclusively on a single factor. For instance, we might run a series of test flights with different wingspans, holding all other factors constant. Then we might try a variety of paperclips and other weights, holding all other factors constant. The trouble with such an approach is that it can conceal important *interactions* among factors. Perhaps wingspan is not the key issue at all, but rather the *ratio* of wingspan to body length. If we only vary wingspan at a single body length we might erroneously conclude that, say, a 4.5-inch wingspan is optimal when the best configuration is for the wingspan to be 1.66 times the body length.

If single-factor experiments are inadequate, we must also be aware that multi-factor experiments can become very expensive and time-consuming. Let's suppose that we have brainstormed 10 possible factors affecting helicopter flight duration. One conventional experimental strategy for screening is to test each factor at two different *levels* initially—high and low, large and small, off and on. Thus, we might run some tests with and without paperclips, long and short wings (defined arbitrarily at first), wide and narrow body, and so on. A *full factorial experimental design* allows us to measure the response variable at all possible combinations of all factor levels. In this case a full factorial design would require us to create enough helicopter designs to cover all possible combinations of the ten different factors. That would require $2^{10}$ or 1,024 different designs. If we were to fly each helicopter ten times to estimate mean duration, we'd need to have 10,240 experimental runs!

We are faced with a tradeoff. Testing one factor at a time is logically inadequate, but testing all possible factors at one time can become onerous. One usual strategy for resolving the tradeoff is to test a few factors at a time. In general, a given factor might have several possible *levels* or values. In our example, there are only two possible levels for the paperclip factor: On or off. Of course, if we had three available sizes of paperclip, then we'd have four different levels: None, small, medium, and large.

For our first example, let's consider three factors with two levels each. Specifically, the three factors are:

- Wingspan: the length of the wings. The levels tested are 3 inches and 4 inches.
- Body width: the width of the folded lower portion of the helicopter. The levels tested are 1 inch and 2/3 inch.
- Clip: presence or absence of a paperclip.

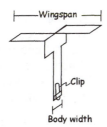

We'll need an experimental design that ensures that we will test all of the combinations of factors and levels. Since we have three factors with two levels each, there are $2^3=8$ combinations in all. We can tabulate the eight possible helicopter models as follows:

| Helicopter | Wingspan | BodyWidth | Paperclip? |
|---|---|---|---|
| A | 3-inch | 1-inch | Yes |
| B | 3-inch | 1-inch | No |
| C | 3-inch | 2/3-inch | Yes |
| D | 3-inch | 2/3-inch | No |
| E | 4-inch | 1-inch | Yes |
| F | 4-inch | 1-inch | No |
| G | 4-inch | 2/3-inch | Yes |
| H | 4-inch | 2/3-inch | No |

Initially, suppose that you will test all eight models yourself, with the assistance of a partner who will time each flight. To gather sufficient data to compute some means and account for measurement error, you'll test each helicopter design eight times.

Why would we replicate our test runs? If we were sure that we could perform the ultimate and perfect test run, then there would be no need to test each model more than once. Replication is a way to accommodate *measurement error*, and to control its effects. Moreover, it gives us a way to estimate the magnitude of measurement error.

One might be tempted to fly model A eight times, then model B eight times, and so on. However, this might compromise the validity of any conclusions we will draw about the models.

1.  *If we conducted the experiment as just described, what problems would you foresee? Why might we question the results of such an experiment?*

Our plan was to have 8 x 8 = 64 experimental runs. A better approach would be to *randomize* the sequence of runs. Suppose we were to assign a unique identifying number for all 64 possible runs. Before running any test flights, we could randomly select a number between 1 and 64, and begin with that particular test flight. By repeatedly generating random values, we could cover all 64 possibilities.

We will use Minitab to determine the randomized sequence of runs, and to create a worksheet into which we can later enter our results. Let's have Minitab set up a worksheet reflecting the full factorial design with eight replications, and randomized sequence of runs.

🖰 **Stat ➤ DOE ➤ Factorial ➤ Create Factorial Design...** Choose 3 factors, then click on **Designs....**

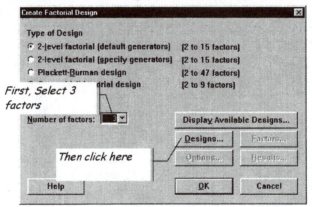

🖰 In the Designs dialog select **Full factorial** and **8** replicates, as shown here (we'll discuss fractional designs, resolution, and center points later) and click **OK:**

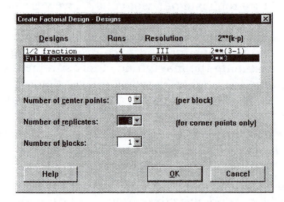

Now click on **Factors...** in the main dialog; this allows us to enter our own factor variable names, replacing the default factor labels A, B, and C. Complete the dialog as shown here.

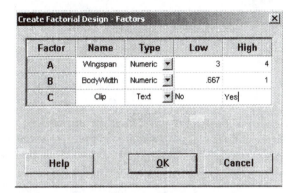

By default, Minitab denotes the two factor levels at –1 and 1. It might refer to them as low and high, yes or no, or we could enter values like 3 and 4 to represent wingspan. The role of the labels is simply to differentiate between the two levels of each factor.

In the main dialog, now click **Options....** For this example, though we want to randomize the sequence of experimental runs, it will be helpful for us to control Minitab's random number generator so that you may follow the example more readily. We can do so by specifying a *base* value for the data generator; enter the value **1316** as shown in the Options dialog box, and click **OK**.

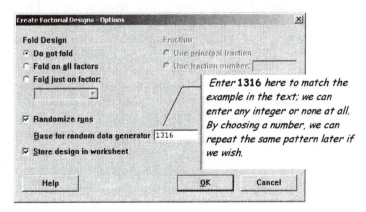

🖰 Now click **OK** in the main dialog. Note that there is now output in the Session Window, and a new worksheet also appears.

---

**Full Factorial Design**

```
Factors:         3   Base Design:         3, 8
Runs:           64   Replicates:             8
Blocks:          1   Center pts (total):     0

All terms are free from aliasing
```
We'll *define* aliasing *later.*

---

2. *Look at the full factorial design summary in the Session Window. Explain, in your own words, what the 3 factors, 64 runs, and 8 replicates are in this experimental design.*

| ↓ | C1 | C2 | C3 | C4 | C5 | C6 | C7-T | C8 |
|---|---|---|---|---|---|---|---|---|
|  | StdOrder | RunOrder | CenterPt | Blocks | Wingspan | BodyWiidth | Cliip |  |
| 1 | 59 | 1 | 1 | 1 | 3 | 1.000 | No |  |
| 2 | 14 | 2 | 1 | 1 | 4 | 0.667 | Yes |  |
| 3 | 41 | 3 | 1 | 1 | 3 | 0.667 | No |  |
| 4 | 28 | 4 | 1 | 1 | 4 | 1.000 | No |  |
| 5 | 13 | 5 | 1 | 1 | 3 | 0.667 | Yes |  |
| 6 | 21 | 6 | 1 | 1 | 3 | 0.667 | Yes |  |
| 7 | 44 | 7 | 1 | 1 | 4 | 1.000 | No |  |
| 8 | 46 | 8 | 1 | 1 | 4 | 0.667 | Yes |  |
| 9 | 23 | 9 | 1 | 1 | 3 | 1.000 | Yes |  |
| 10 | 42 | 10 | 1 | 1 | 4 | 0.667 | No |  |

The first few rows of the resulting worksheet is shown above. The variable **RunOrder** shows the sequence of test flights; **StdOrder** refers to the non-randomized initial sequence. Thus, the plan now is that our first experimental run would have been #59 in our initial sequence: we'll fly a helicopter with 3-inch wings, a 1-inch body, and no paper clip. The next flight will be a 4-inch wingspan, 2/3-inch body, and with a clip.

This particular design forms the basis of an extended example in Session 15. If you have not already done that session, it might be wise to pause here and do Session 15 at this time.

## Blocking

In this example we assumed that a single experimenter performed all of the experimental runs. In many instances this may not be at all practical. Suppose that we wanted to enlist some help to conduct the experimental runs, and that we found three other people to lend a hand.

We would then have four different individuals (or teams) dropping the helicopters and timing the flights.

3.  ***What are some of the complications introduced by having different people involved? List some of the potential "nuisance" factors that this might add to our list of three experimental factors.***

We need to give some thought to the matter of dividing the labor here so as to control for the nuisance factors. We do this by *blocking* the experimental runs. A block is a group of runs that we expect to be relatively homogenous. When we block, we eventually compare the results of each experimental condition within each block. In this case, it will make sense to block by person, allowing each experimenter to run several different helicopter models.

🖱 **Stat ➤ DOE ➤ Factorial ➤ Create Factorial Design...** We return to this dialog, select 3 Factors, and click **Designs**.

🖱 This time, we still have a full factorial design with eight replicates, but now select 4 blocks, as shown here:

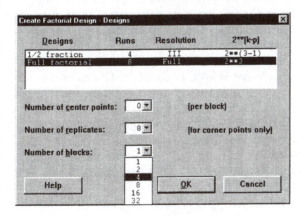

🖱 Click **OK** in this sub-dialog, and then **OK** in the main dialog.

Now look at the new worksheet created by Minitab. You'll notice that there are three factor columns labeled **A**, **B**, and **C**. Look closely at the **Blocks** column. This indicates that particular runs should be assigned randomly to each of the four blocks.

4.  ***How many runs have been assigned to each block?***

**5.** *Examine the factor levels assigned to one of the blocks. How many of the eight possible helicopter designs will be run within that block?*

In this blocked design, we still have 64 runs in all, with eight replicates of the eight possible designs. The analysis of the duration data would proceed as before, with the blocks accounted for.

## Fractional Designs

In this helicopter example, there are few practical limits on the number of experimental runs that we performed. Each run requires a few seconds, and paper is cheap. This is not the case for all experiments, particularly those with more factors or levels, or those with time constraints or with scarce materials.

Suppose we could not afford to run 64 trials in this experiment. One obvious alternative would be to reduce the number of replicates, but this would mean a loss of precision. Another alternative would be to assign only some models to different teams, but that could confound the effects of the teams and the factors. Still another alternative is to create a *fractional* or *incomplete* design in which some combinations of factors and levels are tested explicitly, but others are not. In particular, we'll look at a one-half fractional design in which only half of the possible combinations are tested.

Fractional designs represent yet another tradeoff: by reducing the number of runs we save scarce resources. On the other hand, they introduce some ambiguity, because we don't actually test all possible interactions. We will collect less information from a fractional design than we would from a full factorial design. If resource constraints force us to select the fractional design, we can still obtain useful data that we can analyze fruitfully.

Let's do an example, and then see how the half-fraction design works. We'll assume we still have three factors and four blocks, but will only generate 32 of the 64 possible experimental runs.

✎ **Stat ➤ DOE ➤ Factorial ➤ Create Factorial Design...** By now, you are familiar with this dialog. Select 3 Factors, and click **Designs**.

✎ This time, choose the **1/2 fraction** design at the top of the dialog box. Click **OK** here and then **OK** in the main dialog.

In the Session Window, notice that there is now some additional output, as follows:

---

**Fractional Factorial Design**

```
Factors:   3   Base Design:      3, 4   Resolution with blocks: III
Runs:     32   Replicates:          8   Fraction:  1/2
Blocks:    4   Center pts (total):  0

*** NOTE *** Some main effects are confounded with two-way interactions

Design Generators:   C = AB

Alias Structure

I + ABC

A + BC
B + AC
C + AB
```

---

Notice that there are now 32 runs, still with 3 factors, 8 replicates, and 4 blocks. The output now indicates that the *Resolution with blocks* is III, that "Some main effects are confounded with two-way interactions," that C = AB, and that there is an *Alias Structure*. Let's sort out these new terms after we look closely at the factor combinations that we will run.

Look at the Data Window, and focus on the first eight rows. These are the experimental runs to be conducted by experimenter number 4.

6. ***How many different combinations will this person test? Which specific combinations are they?***

7. ***Now look at the combinations assigned to experimenter number 1. How do they compare to the combinations assigned to experimenter number 4?***

Close inspection of the design reveals that only four of the eight possible combinations of factor levels appear in the list. As shown on the next page we can picture the full-factorial design as a cube, height, width and depth corresponding to the three factors.[2] In this fractional design, four corners (shown in bold) will be tested eight times each. In particular, we will test opposite corners of the cube. Notice that on each face of the cube, two opposite corners are included in the design. Within the four experimental combinations, we have two with high values and two with low values for each of the three factors. This should allow us to get reasonable readings of the main effects of the factors.

---

[2] See Session 15 for a discussion and illustration of the full-factorial design as a cube.

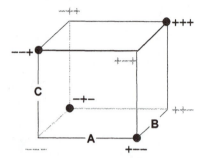

As with any compromise, we gain something and lose something in this half-factorial design. We gain the efficiency by using only half of the possible runs. This comes at the expense of valuable information: because we run only some of the possible combinations, we will not test all possible interactions, and thus some effects will be *confounded* with others. We may not be able to determine whether a particular response value is due to the main effects of the factors or to their interactions. We refer to the confounded main effects and interactions as *aliases*.

In this particular example, the Alias Structure in the Minitab output tells us that we will not be able to distinguish between main effects associated with Factor A and the interaction of factors B and C. In other words, if long-winged helicopters fly longer than others, we won't be able to decide if the long wings cause long flights, or if the interaction of paper clips and body width are the cause. Why? Because we only test long wings with (a) narrow bodies and no clips and (b) wide bodies and clips. We never test long wings, narrow bodies and clips.

Because there will be effects that cannot be clearly identified, we say that this half-fraction design has a weaker *resolution* than the comparable full-factorial design. We say that this is a Resolution III design. Fractional experimental designs commonly are classified as having the following resolutions:

- *Resolution III*: Main effects may be aliased with two-factor interactions, but main effects are not aliased with each other.
- *Resolution IV*: Main effects are not aliased with each other or with any two-factor interactions, but two-factor interactions are aliased with one another.
- *Resolution V*: Neither main effects nor two-factor interactions are aliased with one another, but two-factor interactions are aliased with three-factor interactions.

Since Resolution III designs confound main effects with two-factor interactions, it is wise to restrict their use to instances in which we are confident that there are no substantial interactions. For example, if we had run a full-factorial experiment that showed no interactions, we might economically use a half-fractional design to follow up and test other factor levels.

## Plackett-Burman Designs

If one is reasonably confident that there are no important interaction effects, and therefore a Resolution III design is sufficient, one might want to consider a Plackett-Burman design. These are two-level fractional designs that enable an experimenter to generate data relevant to main effects with relatively few runs. In fact, for a design with $k$ factors Plackett-Burman designs requires a minimum of $k+1$ runs; thus one could design a 7-factor experiment with just 8 runs.

It is beyond the scope of this session to provide a full discussion of these designs. This section provides a very brief introduction to the Plackett-Burman approach. For our example, suppose that we identified five factors for our helicopters, and that from earlier experiments we were satisfied that each has an independent main effect. In our first example in this session, we had three factors and the full-factorial experiment with eight replicates required 64 runs in all. With five factors and eight replicates, we would need $8(2^5) = 256$ runs; suppose further that we can only afford 12 runs this time. Let's use a Plackett-Burman design in this situation.

- Stat ▶ DOE ▶ Factorial ▶ Create Factorial Design...  In this familiar dialog, select Plackett-Burman design with 5 factors.

- In the dialog, click the **Designs...** button and choose 12 runs.

- Now click **Options...** and again enter the value 1316 to see the random number generator.[3]

- Finally, click **Results...**, and select **Summary table and data matrix**, as shown on the next page.

---

[3] Remember that there is nothing special about this value. Choosing a specific seed simply ensures that your results match those shown in this book. In a real experiment, you would presumably choose your own seed value.

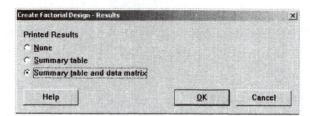

In your Session Window, find the summary of this design (shown below). Note that each of the five factors is tested at its high level in six runs and at its low level in six runs. Each run tests a unique combination of factor levels, and there obviously are factor level combinations that are not tested. In the data matrix of the output, a + represents a run with a high value of a given factor, and a − represents a run with a low value of a given factor.

---

## Plackett - Burman Design

| | | | | |
|---|---|---|---|---|
| Factors: | 5 | Replicates: | 1 | |
| Base runs: | 12 | Total runs: | 12 | |
| Base blocks: | 1 | Total blocks: | 1 | |

Design Matrix (randomized)

| Run | Blk | A | B | C | D | E |
|---|---|---|---|---|---|---|
| 1 | 1 | + | + | − | + | + |
| 2 | 1 | + | + | − | + | − |
| 3 | 1 | + | + | + | − | + |
| 4 | 1 | − | − | − | + | + |
| 5 | 1 | + | − | + | + | − |
| 6 | 1 | − | + | + | − | + |
| 7 | 1 | + | − | − | − | + |
| 8 | 1 | − | + | + | + | − |
| 9 | 1 | − | − | + | + | + |
| 10 | 1 | − | − | − | − | − |
| 11 | 1 | − | + | − | − | − |
| 12 | 1 | + | − | + | − | − |

---

8.  ***In this design, there is no run with all five factors set to their high levels. Identify two more factor level combinations that do not appear in this design.***

Notice also that factors A and B are both high (+  +) in three runs and low in three runs.

9. *In the 12 runs, how often is A high and B low? B low and A high?*

10. *Choose another pair of factors. Does this pair follow a similar pattern of + +, + −, and − − as the AB pair? What does this suggest to you about this particular design?*

From this point on, one would proceed as before—conducting the experiment, recording the data, and analyzing the results. It is important to remember that this design will only reveal main effects, and that we cannot study interactions in a Plackett-Burman design. Therefore, these designs must be implemented with care and with confidence that there are no important interactions.

## Further Reading

Box, G.E.P., Hunter, W.G., & Hunter, J.S. (1978) *Statistics for experimenters*. New York: Wiley (Wiley Interscience).

Hoerl, R.W. & Snee, R.D. (2002) *Statistical thinking: Improving business performance*. Pacific Grove CA: Duxbury.

Montgomery, D.C. (1991) *Design and analysis of experiments*. 3rd ed. New York: Wiley.

Plackett, R.L. & Burman, J.P. (1946) The design of optimum multifactorial experiments. *Biometrika 33*(4), 305–325.

Zikmund, W.G. (2001) *Exploring Marketing Research*. 7th ed. Orlando FL: Dryden.

## Moving On...

It is in the nature of this session that we cannot do exercises with the datasets provided with this text. All of these questions call upon you to design experiments to satisfy particular situations.

Now use the commands illustrated in this session to answer these questions.

1. Staying with the same three factors as in the early section of the session, propose a third *factor level* to test for each of the three factors in a follow-up experiment.

2. If there were three factors, each with three levels, and we ran eight replicates, how many runs would we need in all?

3. Design a full factorial experiment with four factors, each with two levels, ten replicates and no blocking. Describe the key attributes of this experimental design.

4. Design a half-factorial design for the experiment referred to in question #3. Describe the tradeoffs involved in using this design rather than the full factorial design.

5. Design a full factorial experiment with four factors, ten replicates, and 5 blocks. Again, describe the key attributes of this design.

6. Design a half-factorial design for this experiment, also using 5 blocks. Compare the resulting design to that of question #5.

7. Create a Plackett-Burman design for this four-factor experiment (again with 10 replicates and 5 blocks). Compare the resulting design to the two previous designs.

# Session 23

## Methods for Quality

### Objectives

In this session, you will learn to:
- Create and interpret a mean control chart
- Create and interpret a range control chart
- Create and interpret a standard deviation control chart
- Create and interpret a proportion control chart
- Create and interpret a Pareto chart

### Processes and Variation

We can think of any organizational or natural system as being engaged in *processes*, or series of steps that produce an outcome. In organizations, goods and services are the products of processes.

In many processes one dimension of product or service quality may be the minimization of process variation. That is to say, one difference between goods of higher or lesser quality often is their *consistency*. People who are responsible for overseeing a process need tools for detecting and responding to variation in a process.

Of course, some variation may be irreducible, or at times even desirable. However if variation arises from the deterioration of a system, or from important changes in the operating environment of a system, then some intervention or action may be appropriate.

It is crucial that managers intervene when variation represents a problem, but that they avoid unnecessary interventions that either do harm or do no good. Fortunately, there are methods that can help a manager discriminate between such situations.

This session begins with a group of statistical tools known as *Control Charts.* A control chart is a time series plot of a sample statistic. Think of a control chart as a series of hypothesis tests, testing the null hypothesis that a process in "under control."

How do we define "under control?" We will distinguish between two sources or underlying causes of variation:

- *Common cause* (also called *random* or *chance*): Typically due to the interplay of factors within or impinging upon a process. Over a period of time, they tend to "cancel" each other out, but may lead to noticeable variation between successive samples. Common cause variation is always present.

- *Assignable cause* (also called *special* or *systematic*): Due to a particular influence or event, often one which arises "outside" of the process.

A process is "under control" or "in statistical control" when all of the variation is of the common cause variety. Managers generally should intervene in a process with assignable cause variation. Control charts are useful in helping us to detect assignable cause variation.

## Charting a Process Mean

In many processes, we are dealing with a measurable quantitative outcome. Our first gauge of process stability will be the sample mean, $\bar{x}$. Consider what happens when we draw a sample from a process that is under control, subject only to common cause variation. For each sample observation, we can imagine that our measurement is equal to the true (but unknown) process mean, $\mu$, plus or minus a small amount due to common causes. In a sample of $n$ observations, we'll find a sample mean $\bar{x}$. The next sample will have a slightly different mean, but assuming that the process is under control, the sample means should fluctuate near $\mu$. In fact, we will rarely see a sample mean that fluctuate beyond three standard errors of the true population mean.

An $\bar{x}$-chart is an ongoing record of sample means, showing the historical (or presumed) process mean value, as well as two lines representing "control limits." The control limits indicate the region approximately within three standard errors of the mean. Minitab computes the control limits, described further below, and can perform a series of tests to help identify assignable causes. These tests are reliable when our data are approximately normal. An example will illustrate.

Recall the household utility data (in worksheet **Utility**). In this file we have 138 monthly readings of electricity and natural gas consumption in my home, as well as monthly temperature and climate data. We'll start by creating a control chart for the monthly electricity consumption, which varies as the result of seasonal changes and family activity.

We added a room and made some changes to the house, beginning roughly five years along in the dataset. In month 120, we added central air conditioning and upgraded to a new, more efficient furnace. We suspect that the construction project and the presence of additional living space may have increased average monthly electricity usage, and we surely expect the air conditioning to add to electricity use.

**Stat ➤ Control Charts ➤ Variables Charts for Subgroups ➤ Xbar...**
We have one electricity usage measurement per month, stored in **KWHpDay**. We will "sample" three months at a time, representing the four seasons within a year. In the dialog box, select the variable **KWHpDay**, and indicate that the 'subgroup size' is 3 months (as shown below). Click on **Xbar Options**....

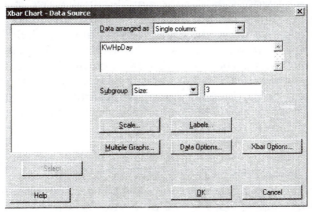

Minitab will perform a series of tests designed to identify evidence of assignable (or special) causes. Click on the **Tests** tab; the eight available tests are listed in the dialog box. Select **Perform all tests**, and click **OK** create the chart.

The dialog box describes the tests quite fully. Each test represents a reason to suspect that the sample means may not be random. In the resulting control chart the points marked with red stars and numbered indicate places at which a specific test failed. Let's take a close look at this control chart.

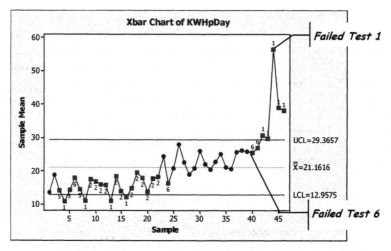

On the horizontal axis is the sample number; we have 27 samples of 3 observations each, collected over time. The vertical axis represents the means of each sample. The green horizontal line, labeled **Mean**, is the grand mean of all samples. The two red lines are the Upper and Lower Control (sigma) Limits. The upper line is 3 standard errors above the mean, and the lower is 3 standard errors below. By default, Minitab estimates sigma using the pooled standard deviation from the sample data. If there are any missing observations in the data, the control limits will vary from sample to sample, and the tests for special causes will *not* be performed.

What does this chart tell us about electricity consumption? For the first 20 samples, the process appears to be in control, except that all of the observations consistently fall below the grand mean. Note that at several of these points, tests 1, 2, 5 and 6 are indicated.

1.  *Why would test #2 indicate a process out of control?*

2.  *Explain what tests 5 and 6 mean, and why they might indicate a process out of control.*

Beginning with sample 21 (month 63) and running until about sample 40, we see a change in this pattern. The red squares and numbers are much less prevalent, but then we see a marked upward shift in electricity usage from sample 40 onward.

3.  *How does this control chart correspond to what you read about the Utility data a few pages ago? Explain with reference to that original story.*

This example illustrates the presence of assignable cause variation. The Xbar chart and tests presume a constant process standard deviation. Before drawing conclusions based on this chart, we should examine the standard deviations among the samples. After all, sample variation is another aspect of process stability. Later we'll illustrate the use of a standard deviation S chart; the S chart is appropriate for samples with 5 or more observations. For smaller samples, such as we have here, the better tool is the Range R chart.

## Charting a Process Range

The Range chart tracks the sample ranges (maximum minus minimum) for each sample. It displays a mean range for the entire dataset, and control limits computed based upon the mean sample range. The range chart has four tests available for assignable causes.

🖱 **Stat ➤ Control Charts ➤ Variables Charts for Subgroups ➤ R...** Just as with the Xbar chart, our data are in a single column. The variable is still **KWHpDay**, and the subgroup size is still 3.

🖱 Select **R Options...**, select the **Tests** tab, and again choose **Perform all tests**.

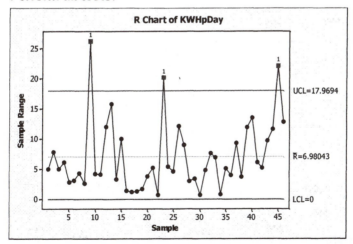

This chart is comparable in structure to the Xbar chart. Note that the lower control limit is closer to the centerline than the upper limit, because the sample range can never be less then 0. In a stable process, the sample ranges should be randomly distributed within the control limits. In three samples (9, 23, and 45) the sample range is more than 3 standard errors above the average sample range. This suggests some instability in the process. Something unusual in those sample periods led to an extraordinary difference between the high and low readings.

It is also important to compare the Xbar and R charts. When a process is under control, both charts should fluctuate randomly within control limits, and should not display any obvious connections (e.g. high means corresponding to high ranges). In practice, we often place the mean and range charts on a single page, as follows.

🖰 **Stat ➤ Control Charts ➤ Variables Charts for Subgroups ➤ Xbar-R...**
Proceed as you did in the prior dialogs.

This Xbar chart is slightly different than before. Here, the standard error is estimated based upon the sample ranges rather than the sample standard deviations. This locates the control limits differently. In this output, we have further evidence of a process out of control at several points, suggesting that the home owner might want to intercede to stabilize electricity use.

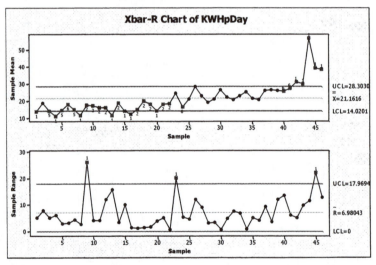

## *An Environmental Example*

In the previous example, the observations were in a single column. Sometimes sample data is organized so that the observations within a single sample are in a row across the worksheet, and each row represents the set of sample observations. To construct control charts from such data, we need only make one change in the relevant dialogs.

To illustrate, open the worksheet called **LondonNO**. This file contains hourly measurements of nitric oxide (NO) concentrations in the air in West London, England. Each row represents one day between January 1 and June 30, 2000 and each column a single hour of the day. Each value in the worksheet is the measured quantity of NO, in parts per billion. A variety of factors presumably influence the amount of NO in the air at one time; thus we may conceive of these figures as arising from an ongoing process.

🖱 **Stat ➤ Control Charts ➤ Variables Charts for Subgroups ➤ Xbar-S…** Since we have 24 observations per sample, let's use the standard deviation as the basis of sample variation. In the drop down list marked **Data are arranged as**, select **Subgroups across rows of.** Move the cursor to the white box below that label, and click once.

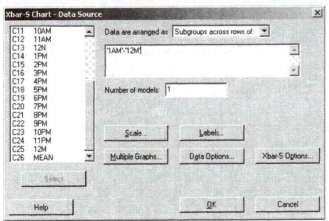

🖱 Then move the cursor into the list of variable names; click and drag to highlight the list of names from **1AM** to **12M** Release the mouse button, and click **Select**.

🖱 Since atmospheric conditions might be seasonal, let's just look at the daily results for the month of January. Select **Data Options…** and indicate that we want to include just rows 1 through 31. Do this by clicking on the **Row numbers** button at the bottom of the dialog list, and typing **1:31** in the white box (see next page).

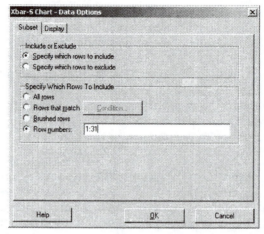

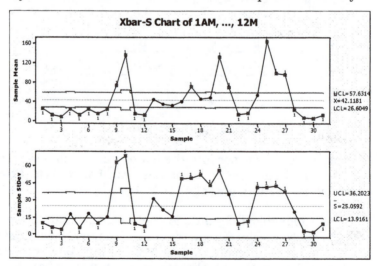

Request all eight tests (under **Xbar-S Options…**, **Tests** tab), and produce the chart. It should look like the graph on the next page.

Here we see many test failures both in the means and the standard deviations. Statistically, this process is not in control. The mean daily concentration of NO varies in an unpredictable way.

## *Charting a Process Proportion*

The previous examples dealt with measurable process characteristics. Sometimes, we may be interested in tracking a qualitative attribute or event in a process, and therefore focus our attention on the *frequency* of that attribute. In such cases, we need a control chart for the sample proportion. For our example, let's consider the process of doing research on the World Wide Web.

There are a number of search engines available to facilitate Internet research. The user enters a search phrase and the engine produces a list of relevant Universal Resource Locators (URL's), or Web addresses. Sometimes, a URL in the search engine database no longer points to a valid Web site. In a rapidly changing environment like the Internet, it is common for Web sites to be temporarily unavailable, move, or vanish. This can be frustrating whether one is conducting research, job-hunting, or even shopping.

One popular search engine is Yahoo!®. Yahoo! offers a feature called the Random Yahoo! Link. When you select this link, one URL is randomly selected from a massive database, and you are connected with that URL. I sampled twenty random links, and recorded the number of times that the link pointed to a site which did not respond, was no longer present, or had moved. I repeated the sampling process twenty times through the course of a single day, and entered the results into the worksheet called **Web**. Open it now.

🖱 **Stat ➤ Control Charts ➤ Attributes Charts ➤ P...** This dialog is similar to those before. Select the **Variable** called **Problems**, and indicate that the **Subgroup size** is **20**. Click **OK**.

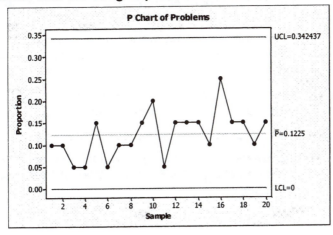

As evidenced in the control chart, this process is under control. All of the variation is within the control limits, and there are no reports of any failed tests. Note that the lower control limit is set at 0, since we can't have a negative proportion of problem URLs. The chart indicates that approximately 12% of the attempted connections encountered a problem of some kind, and that proportion remained stable through the course of a single day.

In this example, all samples were the same size ($n=20$). Had they been of different sizes, we would need a variable indicating the sample size, and would identify that variable in the **Subgroups in** portion of the dialog. When subgroups vary in size, the control limits are different for each sample, and no tests are performed.

## Pareto Charts

The last technique we'll see is known as a Pareto Chart, a tool often used to identify the most frequent causes of a problem or defect. Consider the situation of the Bridgestone/Firestone tire company during the summer of 2000. Within a few short weeks, the United States experienced a large number of serious automobile accidents involving vehicles with Bridgestone/Firestone tires. Reports of these accidents came to the National Highway Safety Administration (NHSA), which compiles data on such events. Although we now know a good deal about the problems with these tires and the conditions under which they failed, in August 2000 this information was not known.

Open the worksheet file called **Tires**. This worksheet contains detailed information about each of the 71 tire-failure accidents reported to the NHSA in July 2000.

Among the variables recorded is the name of the state in which the accident occurred. This is a categorical variable, and thus a simple tally can reveal where these accidents tended to happen.

**Stat ➤ Tables ➤ Tally Individual Variables...** Select the variable **STATE**, and request **Counts** and **Percents**.

4. *In which state did the most accidents occur?*

5. *Five states accounted for nearly 90% of all accident reports to the NHSA in July 2000. Which five states were they?*

The tally lists the states in alphabetical order, providing absolute and relative frequencies. The layout and compact size of the tally make it fairly easy to answer the first of these questions, but the second question requires some effort. We could rerun the tally to include cumulative frequencies, but then we would want the states to appear in order of descending frequency rather than in alphabetical order.

A Pareto chart is a graphical representation of both the absolute frequency and the cumulative relative frequency of a categorical variable. Fundamentally, it is a bar chart in which the category axis is ordered from most to least frequent. In addition, an ogive is superimposed on the bar chart. Let's create a Pareto chart for this variable. Although the state variable does not represent a type of defect, it does identify the location of the accident; Minitab presumes that we are charting defect data.

🖰 **Stat ➤ Quality Tools ➤ Pareto Chart...** Select the variable **STATE**, as shown in the following dialog box.

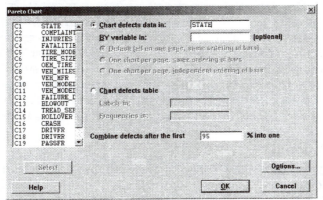

Note that this dialog permits several options, but we'll only make use of the defaults. In the graph below, note also that there are two vertical scales. On the left, we see the simple frequency of accidents in each state. On the right side, we have cumulative percentages. Thus we can see, for instance, that Florida and California accounted for approximately 80% of all accident reports. Consider your response to the previous session question, and note how easy it would be to answer it using this graph.

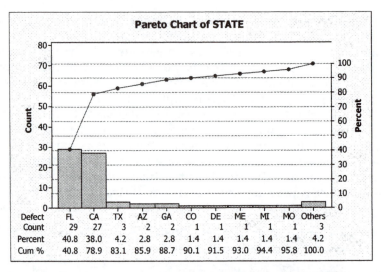

**Pareto Chart of STATE**

| Defect | FL | CA | TX | AZ | GA | CO | DE | ME | MI | MO | Others |
|--------|------|------|------|------|------|------|------|------|------|------|--------|
| Count | 29 | 27 | 3 | 2 | 2 | 1 | 1 | 1 | 1 | 1 | 3 |
| Percent | 40.8 | 38.0 | 4.2 | 2.8 | 2.8 | 1.4 | 1.4 | 1.4 | 1.4 | 1.4 | 4.2 |
| Cum % | 40.8 | 78.9 | 83.1 | 85.9 | 88.7 | 90.1 | 91.5 | 93.0 | 94.4 | 95.8 | 100.0 |

In a production process like tire manufacturing, it is important that manufacturers can trace the root causes of defects and correct them. Sometimes, there may be more than one type of defect or root cause. Assuming that firms cannot attack all problems simultaneously, it is sensible to set priorities and initially deal with problems that are most widespread or serious. Pareto charts can be a very helpful tool in identifying those defects or complaints that are most common.

## Further Reading

Brassard, M. & Ritter, D. (1994) *The memory jogger II*. Methuen MA: GOAL/QPC.

Hoerl, R.W. & Snee, R.D. (2002) *Statistical thinking: Improving business performance*. Pacific Grove CA: Duxbury.

Juran, J.M. & Gryna, F.M. (1988) *Quality control handbook* (4th ed.). New York: McGraw-Hill.

Wadsworth, H., Stevens, K., & Godfrey, B. (1986). *Modern methods for quality control and improvement*. New York: Wiley.

## Moving On...

Use the techniques presented in this session to examine the processes described below. Construct appropriate control charts, and indicate whether the process appears to be in control or not. If not, speculate about the possible assignable causes which might account for the patterns you see.

## Web

1. You can repeat my experiment with the Random Yahoo! Link, if you have access to a Web browser. In your browser, establish a bookmark to this URL:

   http://random.yahoo.com/bin/ryl

   Then, each time you select that bookmark, a random URL will be selected, and your browser will attempt to connect you. Tally the number of problems you encounter per 20 attempts. Repeat the sampling process until you have sufficient data to construct a P chart. NOTE: This process can be very time-consuming, so plan ahead.

   Comment on how your P chart compares to mine.

## Labor1

2. The variable called **PartTeen** is the mean labor force participation rate for teenagers during the observation month. Using a subgroup size of 12 months, develop appropriate control charts to see if variation in teen labor force participation was largely common cause variation.

3. Perform a similar analysis for female labor force participation (**PartF**). Comment on your findings, and speculate about underlying causes.

4. The variable called **HoursM** represents the mean weekly hours worked by manufacturing workers in the observation month. Using a subgroup size of 12 months, develop appropriate control charts to see whether the factors affecting weekly hours were largely of the common cause variety.

5. Using the Calculator, create a new variable called **EmpRate**, equal to total Civilian Employment divided by Labor Force. This will represent the *proportion* of the labor force which was actually employed during the observation month. Using a subgroup size of 12 months, develop appropriate control charts to see whether the factors

affecting employment were largely of the common cause variety.

## LondonCO

The National Environmental Technical Center in Great Britain continuously monitors air quality at many locations. Using an automated system, the Center gathers hourly readings of various gases and particulates, twenty-four hours every day. The worksheet called **LondonCO** contains the hourly measurements of carbon monoxide (CO) for the year 2000, recorded at a West London sensor.

6. Create an Xbar Chart to display the sample means for these data. Note that the data set is organized across the rows of the variables called **1AM** through **12M**. Would you say that this natural process is under control?

7. Do the sample ranges and sample standard deviations appear to be under control?

8. Based on 1995 data, the historical mean CO level was 0.669, and the historical sigma was 0.598. Rerun the Xbar chart, this time relying these historical values. To do so, in the main dialog click the button marked **Xbar options**, and choose the tab marked **Parameters**. In this subdialog, enter the values 0.669 and 0.598 for the historic mean and standard deviation. Does the process appear to be under control? What might account for the appearance of this graph comparing 2000 data to 1995 values?

## Tires

9. When the accident reports began to accumulate, Bridgestone/Firestone needed to know which tires seemed to have the problems. Create a Pareto chart for the variable called **Tire_Model** and describe your findings. Note that the some tire models may have been entered under multiple names by the NHSA.

10. Use a Pareto chart to identify the automobile manufacturers whose vehicles were most often involved in these accidents. Comment on your findings.

## Eximport

11. Using the Calculator, create a new variable called **Ratio**, equal to total exports excluding military aid (**ExnoMA**) divided by general imports (**GenImp**). Using a subgroup size of 12 months, develop appropriate control charts to see whether the factors affecting the ratio of exports to imports were largely of the common cause variety.

## Utility

12. Develop appropriate control charts to analyze the variable called **Therms**, which measures total monthly consumption of natural gas in the household. Use a subgroup size of 12 months, and comment on evidence of assignable cause and common cause variation.

13. Perform a similar analysis of the variable **HDD** (Heating Degree Days). This variable, calculated by the National Weather Service, represents the extent to which outdoor temperatures were low enough to call for artificial heat. Thus, HDD represents a natural on-going process.

## Helpdesk

This worksheet contains a log of inquiries received by a college computer center help desk during the first four weeks of the fall semester.

14. The variable called **Problem** is a brief description of the user's problem. Create a Pareto chart for this variable, and discuss the managerial implications of the results.

15. The help desk manager had hoped that during the first four weeks, both users and help desk assistants would become more adept at their tasks, so that the time required to resolve help inquiries would decline over the first month. Create a control chart to see if that occurred; use a subgroup size of 4.

326

# Dataset Descriptions

The datasets used in each session are available on the CD-ROM accompanying the book, and at the Duxbury Press website in the Data Library. Each of the worksheet files listed here contains detailed descriptions of each variable, as well as citations of the original data sources. This appendix lists all of the datasets, and explains how to locate the detailed descriptions.

## Finding Data Descriptions

To illustrate the documentation available for each dataset, we'll inspect the contents of the worksheet **GSSEduc**, one of six worksheets extracted from the 1998 General Social Survey. This worksheet contains several variables pertaining to the educational experience of the survey respondents.

- **File ➤ Open Worksheet…** Select the file **GSSEduc**.

- **Window ➤ Project Manager** In the left pane of the Project Manager window, highlight the folder marked **GSSEduc**, and you will see summary information about the file in the right-hand pane, as shown on the next page. On your screen, the Location of the file will differ from that shown here; it will refer to the location of this data file on your system.

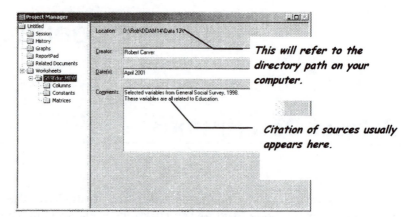

In addition to the general information provided in this screen, you can also obtain detailed descriptions of each variable in the worksheet.

🖰 Move your cursor to the folder marked **Columns**, and click once.

The Project Manager window will now look like this:

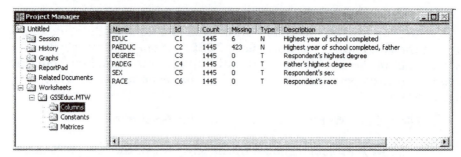

For each variable in the worksheet, you now see the location of the variable, the number of observations, the number of missing observations, a variable type code, and a short description of the variable. The data type codes in Minitab are N for numeric data, T for text (alphanumeric data), and D for dates.

You can also find the worksheet and variable descriptions by moving your cursor into the worksheet, and hovering over the small red triangle in the upper right corner of the column heading cells. The figure on the next page illustrates this feature.

## The Datasets

This is a complete list of the 67 worksheet files used in this book. Those shown with a bullet (•) are new to this edition; the others appeared in *Doing Data Analysis with Minitab 12*.

| Dataset | Short Description |
|---|---|
| Anscombe | Artficial regression data |
| BodyFat | Body measurements |
| • Bowling | Bowling scores |
| • California | Almanac data |
| • Catalog1 | Mail order results |
| • Catalog2 | Mail order results |
| Cholest | Cholesterol measurements |
| • CPSVoting | Current Population Survey data about voting |
| • Eagles | Eagle population estimates |
| Eximport | Export/import data |
| • Faithful | Old Faithful eruptions |
| Falcon | DDT residues in falcon eggs |
| • Florida | Almanac data |
| • Florida Votes | Results of 2000 elections |
| Galileo | Gravity experiments |
| • Gosset | Crop yields |
| • Grades | Student grades |
| • GSSEduc | General Social Survey: education |
| • GSSGeneral | General Social Survey: general demographics |
| • GSSHousehold | General Social Survey: household characteristics |
| • GSSRelig | General Social Survey: religion |

| Dataset | Short Description |
|---|---|
| • GSSSex1 | General Social Survey: sexual behavior |
| • GSSSex2 | General Social Survey: sexual behavior |
| Haircut | Haircut prices |
| • Helicopters | Paper helicopter experiment |
| • Helpdesk | Calls to college helpdesk |
| • Hotdog | Hot dog contents |
| • Illinois | Almanac data |
| • Impeach | Pres. Clinton impeachment votes |
| • Japan Crime | Crime in Japan |
| • JFKLAX | Flight delays between NY and LA |
| Labor1 | Labor market time series |
| Labor2 | Labor market time series |
| LondonCO | Carbon dioxide concentrations |
| • LondonNO | Nitrous oxide concentrations |
| Marathon | Boston marathon results |
| • MCASELA | Massachusetts Student test scores |
| Mendel | Gregor Mendel's pea experiment (summary) |
| MFT | Major Field test results |
| • Michelson | Speed of light measurements |
| • Michigan | Almanac data |
| • Milgram | S. Milgram's experiments on obedience |
| • New York | Almanac data |
| • Ohio | Almanac data |
| • Ohio Votes | Results of 2000 elections |
| Output | Industrial output time series |
| Pennies | Penny tossing results |
| • Pennsylvania | Almanac data |
| • San Diego Crime | Crime in San Diego |
| Sleep | Sleep habits of mammals |
| • StateTrans | State transportation-related measures |
| • Stepping | Heart rate and physical activity |
| Student | Student survey information |
| Swimmer1 | High school swim team results |
| Swimmer2 | High school swim team results |
| • Terrorism | Annual terrorism incidents |
| • Texas | Almanac data |
| • Texas Geog | Almanac data |
| • Texas Votes | Results of 2000 elections |
| • Tires | Tire-related auto fatalities (Firestone) |
| • Tirewear | Tire wear comparisons |

| Dataset | Short Description |
|---|---|
| • Triathlon | Olympic triathlon results |
| US | US economic/demographic data |
| Utility | Household utility usage |
| • Violations | Traffic violations on college campus |
| WATER | US Water usage |
| Web | Errors encountered searching for web sites |

# Working with Files

## Objectives

This Appendix explains several common types of files that Minitab supports and uses. Though you may not use every kind of file, it will be helpful to understand the distinctions among them. Each file type is identified by a three-character extension (like MTW or TXT) to help distinguish among them. For those just getting started with statistics and Minitab, the most useful file types are these:

| Extension | File Type |
|-----------|-----------|
| MTW | Minitab Worksheet |
| MTP | Minitab Portable Worksheet |
| TXT | Text file for Session, History, or Data |
| MGF | Minitab Graphics |
| MPJ | Minitab Project |

The following sections review these types of files, and explain their use. In addition, there is a section which illustrates how you can convert data from a spreadsheet into a Minitab worksheet.

## Worksheets

Throughout this book, you have read data from Minitab worksheet files. These files have the extension MTW, and the early exercises explain how to open and save such files. These files just contain raw data (numeric, text, or date/time). In Release 14 of Minitab, additional information is embedded in MTW files, permitting the user to annotate or describe the contents of individual variables and to specify the order of values for text variables, as described in Appendix A.

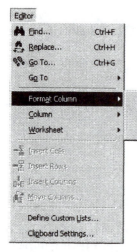

Generally, when you enter data into an active worksheet, the default settings of column format (data type, column width, and so on) are acceptable. Should you wish to customize some of these elements, you'll find relevant commands on the **Editor** menu. For example, as shown to the left, you can format a column as numeric, text, or date data; you can also include a column description with the stored file, or specify the width of a particular column.

It is often useful to specify the order of values for a particular variable. By default, Minitab assumes that values should be in ascending alphabetical or numerical order. However, suppose you have an ordinal variable representing size, and the three possible values are "Small, Medium, and Large." When Minitab computes a Tally or constructs a Chart for the variable, the values will be ordered as "Large, Medium, Small." With the **Editor ➤ Column ➤ Value Order** command, you can enforce whatever order is meaningful.

Some earlier versions of Minitab did not allow all of these features, which is one reason that the **Save Worksheet As** dialog lists several Minitab formats for saving data files (as shown here):

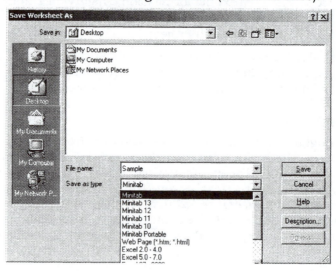

Also note that data in a worksheet can be saved as a Minitab Portable file, or in one of several popular spreadsheet or database formats. The latter options are discussed later. The "Portable" format is a version-independent format for earlier versions of Minitab. If you need to save a worksheet for other Minitab users, but aren't sure which version of Minitab they use, this is the safe choice of file formats.

Note that when you save a Worksheet file, you are *only* saving the data that you entered or computed during a given session. If you also want to save the results of analysis, or the commands you have given, you must save a Session or History file.

## Session and History Files

After doing analysis with Minitab, you may want to save a record of the work you've done, particularly if you need to complete it at a later time. That is the point of Session and History files. These files are "transcripts" of the output and command (respectively) you have given during a working session. By default, they are 'text files' meaning that they contain no formatting (boldface, etc.), but only contain words and numbers.

Both of these two file types contain everything that you see in the respective Minitab Window. The Session output is familiar enough, but the History file might appear a bit odd. As you select commands and options from various menus and dialogs, Minitab translates these choices into a command language. A History file contains all of your choices, expressed in that language. To save a Session or History file, first select either the Session or History Window, as appropriate, by clicking anywhere in the window. Here is an example, assuming the Session Window is active:

🖱 **File ➤ Save Session Window as...**

Make this selection to save Session output. Minitab automatically appends the TXT suffix, though you could select Rich Text Format, allowing you to retain formatting of the Session Window. Naturally, if a History Window is active, the menu selection refers to History; otherwise, the commands are the same.

## Graph Files

In addition to Session Window output, various commands create graphs. Each graph is an object in its own right, appearing in its own window. Likewise, an active graph can be saved in a file. By default when you save a graph, Minitab stores it as a Minitab Graphics File (MGF extension); such a file can be re-opened, displayed or printed during a later Minitab session (**File ➤ Open Graph**).

Alternatively, you can save graphs in one of several graphics formats (JPG, TIF, BMP) permitting you to exchange a graph object with other application software.

## Minitab Projects

A project file is analogous to a binder in which you might store several related items. A Minitab Project file stores a collection of Data Worksheets, Session, History and Graphics files all in one place.

To create a Project file, you select

🖰 **File ➤ Save Project...** By default, this command saves all open windows. To exclude an item, just close that window.

Similarly, when you open a project (**File ➤ Open Project...**), all of the saved windows and elements re-appear on the screen, allowing you to resume work right where you left off.

## Converting Other Data Files into Minitab Worksheets

Often one might have data stored in a spreadsheet or database file, or want to analyze data downloaded from the Internet. Minitab can easily open many types of files. This section discusses two common scenarios; for other file types, you should consult the extensive Help files provided with Minitab. Though you do not have the data files illustrated here, try to follow these examples with your own files as needed.

### Excel spreadsheets

Suppose you have some data in a Microsoft Excel spreadsheet, and wish to read it into a Minitab worksheet. You may have created the spreadsheet at an earlier time, or downloaded it from the Internet. This example shows how to open the spreadsheet from Minitab.

First, it helps to structure the spreadsheet with variable names in the top row, and reserve each column for a variable. Though not necessary, it does simplify the task. Such a spreadsheet is shown here:

Assuming the spreadsheet has been saved as an Excel (.xls) file, called **Grades**, you would proceed as follows in Minitab:

🖱 **File ➤ Open Worksheet...** In the dialog choose the appropriate directory, and select **Excel (*.xls)** as the file type. You should then see your file listed. Select the desired file name.

🖱 Click the **Options** button in the **Open Worksheet** dialog, and another dialog box will open (as show below).

If information in the spreadsheet is arranged as described earlier (variables in columns, and variable names in the first row) the **Options** are unnecessary. However, if there are no variable names, or if the data begin in "lower" rows of the spreadsheet, this dialog permits you to specify where Minitab will find the data.

Moreover, the **Options** dialog also allows for specific instructions regarding the treatment of missing data. In this example, we have no need of the options.

🖱 Click **OK** on the **Options** dialog, and **OK** on the **Open Worksheet** dialog, and the data from the spreadsheet will be read into the Data Window. You can then analyze it and save it as a Minitab worksheet.

| | C1-T | C2 | C3 | C4 | C5 | C6 | C7 | C8 | C9 |
|---|---|---|---|---|---|---|---|---|---|
| | Student | Quiz1 | Midterm | Quiz2 | Homework | FinalExam | | | |
| 1 | Appel | 97 | 95 | 88 | 95 | 92 | | | |
| 2 | Boyd | 85 | 90 | 88 | 85 | 78 | | | |
| 3 | Chamberlin | 86 | 86 | 82 | 74 | 80 | | | |
| 4 | Drury | 73 | 68 | 70 | 55 | 65 | | | |
| 5 | Elliott | 64 | 70 | 72 | 73 | 78 | | | |
| 6 | Franklin | 82 | 90 | 75 | 80 | 85 | | | |
| 7 | Gooding | 98 | 96 | 94 | 98 | 95 | | | |
| 8 | Harriman | 80 | 82 | 86 | 74 | 88 | | | |
| 9 | Ingram | 100 | 97 | 100 | 100 | 95 | | | |
| 10 | Jenkins | 78 | 71 | 76 | 77 | 75 | | | |
| 11 | Karp | 72 | 60 | 68 | 80 | 77 | | | |

## Data in text files

Much of the data available for downloading from Internet sites is in text files, sometimes referred to as ASCII format.[1] As just described, Minitab can read data from these files, but needs to be told how the data are arranged in the file.

In these files, observations appear in rows or lines of the file. Text files generally distinguish one variable from another either by following a fixed spacing arrangement, or by using a separator or "delimiter" character between values. Thus, with fixed spacing, the first several rows of the student grade data might look like this:

```
Appel        97    95    88    95    92
Boyd         85    90    88    85    78
Chamberlin   86    86    82    74    80
Drury        73    68    70    55    65
Elliott      64    70    72    73    78
Franklin     82    90    75    80    85
```

---

[1] ASCII stands for American Standard Code for Information Interchange. Unlike a Minitab worksheet or other spreadsheet format, an ASCII file contains no formatting information (fonts, etc.), and only contains the characters which make up the data values.

In this arrangement, the variables occupy particular positions along each line. For instance, the student's name is within the first 11 spaces of a given line, and his/her first quiz grade is in positions 12 through 14.

Alternatively, some text files don't space the data evenly, but rather allow each value to vary in length, inserting a pre-specified character (often a comma) between the values, like this:

```
Appel,97, 95, 88, 95, 92
Boyd, 85, 90, 88, 85, 78
Chamberlin, 86, 86, 82, 74, 80
Drury, 73, 68, 70, 55, 65
Elliott, 64, 70, 72, 73, 78
Franklin, 82, 90, 75, 80, 85
```

Logically, both of these lists of data contain all of the same information. To our eyes and minds, it is easy to distinguish that each list represents six variables. Though there is no single "best format" for a text file, it is important that we identify the format to Minitab, so that it correctly interprets what it is reading. To see why this matters, let's assume that the text file contains the data evenly spaced in pre-specified positions (as in the first example).

In the **Open Worksheet** dialog, we select the **Text (*.txt)** file type. Before opening the file, we click on **Preview**, and see:

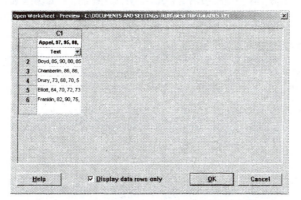

This dialog shows that all of the data will be read into Column C1, and assigned a rather odd name. Clearly, this was not what we had in mind. There are two problems here. First, Minitab is interpreting the first row of data as if it contained variable names. Second, it thinks that each line is one long text variable, rather than six individual variables.

Here is where the **Options** dialog can help. This completed dialog indicates that there are no variable names in this text file, and that the data are not separated by any single special character.

Click **Cancel** in the **Preview** dialog, and then click on **Options** in the **Open Worksheet** dialog.

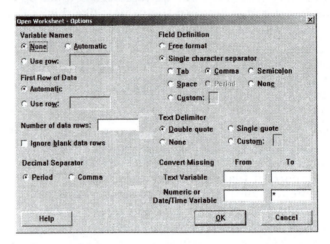

Under **Variable Names**, select **None**, and under **Field Definition**, choose **Single character separator**, and click **Comma**. With the options as shown, a preview of the worksheet now looks like this:

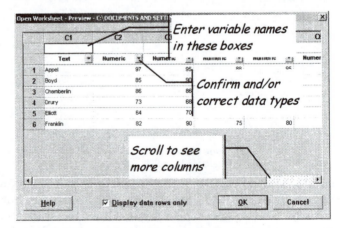

This is what we wanted. By clicking **OK** in the **Preview** and the **Open Worksheet** dialogs, the data will be read into the Data Window as before.

Though this discussion has not covered all possibilities, it does treat several common scenarios. By using the Help system available with your software, and patiently experimenting, you will be able to handle a wide range of data sources.

# Organizing a Worksheet

## *Choices*

Most of the datasets accompanying this book contain a single sample of subjects (states, companies, people, etc.). Within the worksheet, each row contains information about one subject, and each column represents a variable or a characteristic of the subject.

This structure represents one of three fundamental approaches to building a worksheet, and most Minitab commands anticipate (if not require) that the data be organized in this way. Nonetheless, there are times when a different approach might be necessary or desirable. The goals of this section are to help you select a method of organization for data you might collect, and to work more readily with data provided with this book or your textbook.

Consider a small class of 10 students, with the following information concerning each person:

| Name | Sex | Height | Class |
|------|-----|--------|-------|
| Amy | F | 61 | JR |
| Ben | M | 65 | JR |
| Charles | M | 72 | SO |
| Debra | F | 66 | SR |
| Elliott | M | 70 | JR |
| Francine | F | 62 | JR |
| George | M | 74 | SO |
| Hannah | F | 64 | SR |
| Ivan | M | 65 | SR |
| Jaime | F | 66 | JR |

We'll consider three ways to represent these data in a Minitab worksheet. The first, which is most common in our worksheet files, is known as *stacked data* format. Stacked format is the most versatile of the arrangements, permitting virtually any Minitab operation.

## Stacked data

Most Minitab procedures and commands are designed to operate on stacked data. In this arrangement, each column represents a variable or attribute of one subject in the sample, and each row represents one subject. Thus, all of the observations for a given variable are 'stacked up' in a single column. The sample size $n$ equals the number of rows.

| | C1-T | C2-T | C3 | C4-T | C5 | C6 | C7 |
|----|--------|------|--------|-------|----|----|----|
| | Name | Sex | Height | Class | | | |
| 1 | Amy | F | 61 | JR | | | |
| 2 | Ben | M | 65 | JR | | | |
| 3 | Charles | M | 72 | SO | | | |
| 4 | Debra | F | 66 | SR | | | |
| 5 | Elliott | M | 70 | JR | | | |
| 6 | Francine | F | 62 | JR | | | |
| 7 | George | M | 74 | SO | | | |
| 8 | Hannah | F | 64 | SR | | | |
| 9 | Ivan | M | 65 | SR | | | |
| 10 | Jaime | F | 66 | JR | | | |
| 11 | | | | | | | |

Note that we could think of these data as representing samples from two populations (females and males) or three populations (Sophomores, Juniors, and Seniors), depending on the questions and issues under study.

When organized as stacked data, the sub-populations are identified by a variable (Sex or Class). In contrast, *unstacked* data organization assigns observations for a single variable to *different* columns, depending on the sub-population.

## Unstacked data

Suppose that our major interest were in comparing the heights of male and female students. In that case, we might segment our data into two sub-samples, and enter it into the worksheet this way:

| · | C1-T | C2 | C3-T | C4-T | C5 | C6-T | C7 |
|---|------|----|------|------|----|----|----|
|   | Name_F | Height_F | Class_F | Name_M | Height_M | Class_M | |
| 1 | Amy | 61 | JR | Ben | 65 | JR | |
| 2 | Debra | 66 | SR | Charles | 72 | SO | |
| 3 | Francine | 62 | JR | Elliott | 70 | JR | |
| 4 | Hannah | 64 | SR | George | 74 | SO | |
| 5 | Jaime | 66 | JR | Ivan | 65 | SR | |
| 6 | | | | | | | |
| 7 | | | | | | | |
| 8 | | | | | | | |
| 9 | | | | | | | |
| 10 | | | | | | | |
| 11 | | | | | | | |

*(Worksheet 2 \*\*\*)*

Arranged in this way, the Sex variable is omitted, since it is implied by the separation of Names, Heights, and Classes. Logically, this contains all of the information as in the stacked arrangement. Moreover, some Minitab commands (such as the Paired t-test) require unstacked data.

## Summarized data

Sometimes, we begin our work with summary results. For example, we may want to analyze a table published in a journal or news article. We don't actually have access to the original raw data, but instead have only a cross-tabulation or frequency distribution. This limits our analytical options severely, but not completely. One possible summary of our student is as follows:

| C8-T | C9 | C10 | C11 |
|------|------|------|------|
| Year | Female | Male | |
| SO | 0 | 2 | |
| JR | 3 | 2 | |
| SR | 2 | 1 | |

Clearly, we have lost a good deal of information here—we no longer record the names or heights of the individual students. We don't know which three females are sophomores. Nonetheless we could still analyze this table, say, to see if gender and class were independent.

# Index

## Index